THE
Behaviour
OF DOGS AND CATS

written by members of the APBC

Edited by John Fisher

Stanley Paul
London

First published 1993

1 3 5 7 9 10 8 6 4 2

© Association of Pet Behaviour Counsellors 1993

John Fisher has asserted his right under the Copyright, Designs and Patents Act, 1988
to be identified as the editor of this work

First published in the United Kingdom in 1993 by Stanley Paul & Co. Ltd
Random House, 20 Vauxhall Bridge Road, London, SW1V 2SA

Random House Australia (Pty) Limited
20 Alfred Street, Milsons Point, Sydney
New South Wales 2061, Australia

Random House New Zealand Limited
18 Poland Road, Glenfield
Auckland 10, New Zealand

Random House South Africa (Pty) Limited
PO Box 337, Bergvlei, South Africa

Random House UK Limited Reg. No. 954009

A CIP catalogue record for this book is available from the British Library

ISBN 0 09 177816 6

Typeset by SX Composing Ltd, Rayleigh, Essex

Printed in Great Britain by Mackays of Chatham

Contents

About the Authors

Roger Abrantes
Roger Abrantes, ethologist, DF cand.art., APBC, was born in Portugal in 1951. Scientific Director of the Etologisk Institut in Denmark, Roger also advises the Technologic Institute and the Danish Police Force and lectures at the Danish Veterinary University. He is the author of ten books published in Danish, Swedish and Norwegian.

David Appleby
David Appleby has been working professionally with dogs since 1973 when he became an RAF Police Dog Handler serving at many stations in the UK and abroad. In 1981 he joined the Guide Dogs for the Blind Association as a guide dog mobility instructor and since 1986 he has been working as a behavioural counsellor. David has referral clinics throughout the Midlands and Wales, including Cambridge University Veterinary School in conjunction with the Companion Animal Research Group, where he is the visiting behavioural counsellor.

Gwen Bailey
Gwen Bailey graduated from Reading University in 1982 with a BSc (Hons) degree in Zoology (with Physiology and Biochemistry). In 1988 she was appointed Information Officer with the Blue Cross, one of Britain's largest animal welfare charities. For the past three years, Gwen has been successfully solving behaviour problems that occur when a dog has been rehomed by the Society, helping to prevent dogs with behavioural problems from being passed from home to home. She also gives advice to owners who are about to give up their dog for adoption to the Society's shelters, thereby helping them to get rid of

the problem rather than the dog. In 1991, Gwen was appointed full-time Dog Behaviour Therapist, the first to be employed by an animal charity.

Ian Dunbar

Ian Dunbar PhD, BSc, BVetMed, MRCVS, APBC, was born in Hertfordshire and attended the University of London, where he read for degrees in physiology and veterinary science. He subsequently obtained a doctorate in animal behaviour from the University of California. Although he still lives in California, he is a frequent visitor to England where he is in great demand as a speaker.

Ian is regarded by many people as the 'father' of behaviour therapy. In 1979 he published *Dog Behaviour – Why Dogs Do What They Do*, and he followed this with a series of booklets on different aspects of the subject. His research into the beneficial effects of early socialization and training with puppies led to the foundation of his Sirius puppy training system which is widely used in this country.

John Fisher

John Fisher is one of the founder members of the APBC. He has had over twenty years' experience of working with dogs on a professional basis. This includes handling and training operational patrol dogs and specialist task dogs; instructing at pet obedience classes; training and handling dogs for film and television, through to his current full-time occupation of advising pet owners on how to overcome the behavioural problems that they are experiencing with their dogs. Already a successful author on the subject, his books *Think Dog!* and *Why Does My Dog?* have now been followed up with a third book called *Dogwise . . . The Natural Way to Train Your Dog*. John is the current Chairman of the APBC.

Margaret Goddard

Since graduating from the School of Veterinary Science at Bristol University in 1977, Margaret Goddard MRCVS has worked exclusively on the small animal side of veterinary practice and is at present part of the small animal team in a large, mixed practice in Salisbury, Wiltshire.

Problems encountered in her day to day work led to an increasing interest in the behaviour difficulties of companion animals. As a member of the APBC, Margaret hopes to encourage more involvement of her veterinary colleagues in the understanding and treatment of behaviour problems in companion animals.

Sarah Heath

Sarah Heath graduated from Bristol University Veterinary School in 1988 and since then has worked in mixed general veterinary practice at the Croft Veterinary Centre, Brackley, Northants. Sarah was the first Veterinary member of the Association of Pet Behaviour Counsellors and she regularly sees cases of pet behaviour problems both within the practice and on referral from other veterinary surgeons. She is the Veterinary Officer to the APBC and is keen to see more involvement of the veterinary profession in the area of companion animal behaviour therapy and closer association between veterinary surgeons and behaviourists. Sarah writes a regular bimonthly column for the *Veterinary Times* magazine. Her book *Why Does My Cat?* will be published in October 1993.

Anne McBride

Anne McBride has a BSc (Hons) in Psychology, a PhD in Animal Behaviour and is a Visiting Fellow at Southampton University. She currently sees cases in London and Southampton and runs puppy classes as well as lecturing on the assessment and treatment of behavioural problems and early socialization. She has published three books as well as a number of scientific papers. Her special interests in the pet behaviour sphere include the emotional needs of pet owners and the effect on owners of pet loss.

Peter Neville

Peter Neville graduated from the University of Lancaster in 1979 with a BSc (Hons) degree in Biology. He went on to complete postgraduate studies on feral cat, mole and seal behaviour and human/companion animal relations during three years as Research Biologist for the Universities Federation for Animal Welfare. He is currently Visiting Animal Behaviour Therapist at the Department of Veterinary Medicine, Bristol Veterinary School and has been in practice for the referral and treatment of behaviour problems in pets for over seven years. Peter Neville is one of the founder members and current Honorary Secretary of the Association of Pet Behaviour Counsellors. He has recently been awarded a Doctorate DHc by the Etologisk Institut in Denmark for his studies in feline behaviour and the development of behaviour problem treatment techniques.

A frequent broadcaster on national radio and television, Peter is the author of the bestselling books *Do Cats Need Shrinks?* and *Do Dogs Need Shrinks?* He also provides regular features for pet owners in *Dogs Today, Cat World, Wild About Animals* and has a regular column in the *Sun* national newspaper.

Erica Peachey

Erica Peachey BSc (Hons) graduated from Hull University with a degree in Psychology, specializing in Animal Behaviour. She worked as a Research Associate at the Royal (Dick) School of Veterinary Studies, University of Edinburgh, before spending time working with most of the main Animal Behaviour Consultants in this country. Now based in the Wirral, she holds regular clinics all over the country, and also lectures on behavioural problems, treatment and prevention.

Her 'Good Puppy Course' aims to prevent problems occurring, teaching owners and their dogs to have a better understanding of each other. Owners are able to learn about their puppies and also the importance of responsible dog ownership. Puppy parties are held from her clinics, in conjunction with the veterinary surgeon.

Erica is also involved with various rescue organisations and is treasurer of a new society called PETS, which aims to promote humane education and runs a hospital visiting scheme.

John Rogerson

John Rogerson is an internationally acclaimed authority on both training and behaviour and for the last five years has been the most sought after speaker in the United Kingdom, regularly speaking to groups involved in breeding, training, animal welfare and canine assistance. John is regularly consulted by most groups involved in the world of dogs and the techniques that he pioneered have now become standard practice for the majority of people involved in training or behaviour modification. He is the author of four books on behaviour and training techniques.

Robin Walker

After Kimbolton School and national service in the Royal Artillery, Robin Walker BVetMed, MRCVS graduated in 1964 from the Royal Veterinary College at the same time as his wife, Jill, and they have worked together ever since. After a year as House Surgeons at the

RVC, they practised in Maidstone and Belgravia before settling in Worcester where they have a thriving small animal practice. Robin has been veterinary surgeon to West Mercia Police Dog Section since 1982 and has a particular interest in stress, injury and mental disturbance in dogs particularly when caused by aggressive training techniques. Having been a school prefect, regimental policeman, vice-president of a boisterous student union, employer, parent, adult educator (teaching Modern Greek at the local Technical College), and a volunteer assistant at a Special Day Centre for children with emotional and behavioural problems, Robin has strong views about the damage caused to children (and adults) by aggressive parenting and teaching techniques. For light relief he studies Ancient Egyptian, Greek and Roman medical history and delivers yachts to Greece and Turkey.

Ruth Williams

Ruth Williams BVetMed, MRCVS graduated from the Royal Veterinary College in 1974. She worked for several years in mixed practice before deciding to specialize in companion animals in 1985. It was at this stage in her career that she realized just how many pets were being presented for euthanasia because of behavioural problems and her concern prompted an interest in trying to rehabilitate these animals. Ruth became a member of the APBC in 1991 and is at present based in a hospital practice in Gwent.

In her spare time, Ruth breeds, works and shows long-haired Weimaraners. She has competed with considerable success in Working Dog Trials and has won the Weimaraner Club of Great Britain's 'Working Weimaraner Of The Year' award with two of her dogs.

Foreword
Ian Dunbar

I would strongly recommend that readers leave this foreword to last and start reading the text right away. What a truly magnificent book! Straightforward, wonderful to read and overflowing with a wealth of user-friendly, pet-friendly, practical information. Not only is it an honour to be asked to provide the foreword but also it was a thoroughly enjoyable privilege to read the book before anyone else. In fact, the book is so good I could easily write a glowing foreword longer than the book itself. The dogs and cats of this world have waited a long time for this edition. And so, with warm woofs and contented purrs, all our pets join together in offering heartfelt thanks to the members of the Association of Pet Behaviour Counsellors (APBC).

This book will be enjoyed by dog and cat owners alike. It will also be welcomed and valued by veterinary surgeons and animal behaviour counsellors, acknowledged by all pet professionals as a landmark text. Compassion and common sense has at long last returned to the husbandry of pet dogs and cats. Above all, the book's message will be cherished by dogs and cats themselves.

The sheer scope of the authors' cumulative expertise is overwhelming. Indeed, full kudos must go to the editor, John Fisher, who has done an outstanding job bringing together a remarkable group of authors, comprising veterinary surgeons, animal behaviourists, psychologists, biologists, ethologists, zoologists and professional dog trainers. Moreover and notwithstanding the eclectic background of the impressive pack and pride of authors, each chapter champions a common sentiment which runs throughout the entire book: to acknowledge the animals' point of view; to recognize and respect their needs; to prevent problems rather than create them; and to train, or rather educate, our cats and dogs using brain instead of brawn.

Although mainly a book about dog behaviour, feline fanciers will be enthralled by the comprehensive chapter on cat behaviour problems.

Written by Peter Neville, this chapter is worth the price of the book alone.

The first five chapters jointly underscore the crucial importance of early socialization for the prevention of aggression in dogs. Sarah Heath offers a practical veterinary perspective emphasizing puppy training classes as a virtual panacea for potential and/or incipient temperament problems. Without a doubt, the key to the success of preventative puppy programmes lies with veterinary practitioners. Pet behaviour problems have a critical impact on the veterinary profession: unruly pets with tricky temperaments can be time consuming and potentially dangerous to handle and ill-behaved pets seldom remain clients for long, since they are often surrendered to animal shelters. The vet loses a client, the owner loses a companion and the pet often loses its life. Indeed, even a simple house-training problem can be the equivalent of a terminal illness.

Ethologist Roger Abrantes re-emphasizes the need for puppy socialization when he states that the first signs of aggression in puppies occur as early as four or five weeks of age. Although domestic dogs have been selectively bred for ease of socialization with people for over ten thousand years, a domestic dog is not fully domesticated until it has been socialized.

Veterinary surgeon Robin Walker addresses the other most common cause of aggression, namely the owner, or rather *the owner's aggression*, i.e., antiquated, unnecessarily harsh, physically punitive, combative, adversarial and bullying so-called 'training' methods. If you ever want to know why your dog doesn't come when called, look in the mirror – there's your problem. There is absolutely no place for physical or psychological violence in teaching and training. In the words of Confucius: 'There is no need to use an axe to remove a fly from the forehead of a friend.' Robin's own words just about sum it up: 'Harsh training is rather like attempting to alter the shape of slippery bars of soap by treading on them. Some bars may be shaped to your satisfaction, sadly a few will be smashed and a large number will escape from underfoot at great speed and in totally unpredictable directions.'

Citing that 20–25% of pet dogs have behaviour problems and that many fear people, and after reviewing scientific research on socialization, ex-RAF Police Dog Handler and Guide Dog for the Blind instructor David Appleby quickly gets down to the nitty-gritty and offers readers a wealth of sound practical advice on how to socialize puppies to people, especially to strangers and men, and more especially to children. A wonderful chapter – if you have a puppy, read it right away. If you are a veterinary surgeon or dog trainer, the

Things That Can Be Done section of this chapter should be mandatory reading for every new puppy owner.

World renowned dog trainer, lecturer, behaviour counsellor, friend of dogs, and altogether good guy, John Rogerson describes handling and gentling techniques for controlling large dogs. Not surprisingly, he accentuates the intelligent notion of starting in puppyhood. Even so, his gentle but firm, enjoyable but structured techniques are effective with adolescent and adult dogs. John's universally accepted maxim, *control the games and control the dog* by establishing mutual cooperation, produces a dog which is happily compliant and so *wants* to please its owner.

The chapters on *Phobias* and *Separation Anxiety* by Robin Walker and veterinary surgeon Margaret Goddard respectively, both re-emphasize the importance of socialization. Lack of socialization often leaves dogs ill-prepared to cope with stressful circumstances, such as the owner's absence. Similarly, undersocialized and/or sensitive dogs may spontaneously develop phobias when over-reacting to stimuli which are intrinsically a little scarey, such as loud noises.

Blue Cross dog behaviour therapist Gwen Bailey brings light, innovation and success to the otherwise heart-rending and sometimes soul-destroying task of rehoming unwanted dogs – mostly adult dogs with behaviour problems. The story of five-year-old Pepe the Dachshund aggressively guarding the telephone to prevent his 86-year-old owner from talking on the phone (and ignoring the dog) would be comical were it not so sad. Luckily, intervention before surrender enabled Pepe and owner to remain friends and companions.

Education is the key, Gwen's educational syllabus for new owners of successfully rehomed dogs is dictated by the most common reasons pets are surrendered, as such it is both representative and utterly useful – this is the information pet owners need to know.

Following the theme of people education, Erica Peachey's people-problems chapter hits the nail right on the head. It's like Grannie and the Jaguar – no one in their right mind would allow a non-driver to drive a high-speed car. But this is what we do with dog owners all the time. Novice owners with novice dogs – the blind leading the blind. But as Erica notes, guide dogs are trained extensively before being introduced to their new owners and similarly the owners receive instruction before they receive the dog. Perhaps this would be a good format for dog training classes. Written from the owner's viewpoint, Erica's people-training rules should be essential reading for anyone involved with pet owners. Most trainers will readily acknowledge, training dogs is the easy bit, it's training the people which stretches patience and challenges teaching skills.

Editor John Fisher's gem of a chapter celebrates a welcomed transition in dog training. Goodbye to the trainers from the dark side with their clinky-sparky collars, their small hearts and narrow minds, their brittle egos and bullying ways of, 'Tell it!' (even though it doesn't understand), 'Make it!' (even though few owners dare and no child could), and 'Reward it!' Ha! We all know they couldn't praise a moribund worm. Instead welcome to dog-friendly, inducive methods, not only to lure and entice dogs to learn *what* we want, but also to reward dogs so they learn *why* it is in their best interests to respond eagerly and willingly. And welcome to methods which are user-friendly, methods which are easy to master, efficient, effective and above all, enjoyable. By using lures and rewards to increase learning speed and by training the last behaviour in a sequence first, John has eliminated years of dog and owner frustration from many of our training programmes.

The end of the dog's sunset years is the topic of this wonderful book's sunset chapter by animal behaviourist Anne McBride. Touching and mindfully written from the heart, this chapter will mend the troubled hearts and minds of people who are grieving over the loss of a pet. Anne's chapter secures an unconditional reservation for this book on the bookshelf of every veterinary clinic and every animal shelter. Anne's chapter will help people cope with grief, it will also make some people cry. But as Anne says, 'It's all right to cry'. I guess that's all we needed to know.

But loved ones whose loss we may grieve tomorrow are alive and well and living with us today. Whereas no-one can remotely comprehend the full nature and magnitude of the feelings of the bereaved over the loss of a loved one, nearly everyone can recognize, enjoy and benefit from an overt display of love and affection for the living. For those of you who are currently sharing your lives with a happy and healthy pet, now is the time for a celebration of life.

'. . . you'll remain in our hearts, wherever you are,
pulling sleds, burying bones or chasing that star.'
Ode to Omaha, Jean Farquhar

Introduction
Ruth Williams

Tessa's Story
....

Tessa was born in November 1990, one of a litter of seven German Shepherd Dog (GSD) puppies. The litter was reared in a kennel, and, other than at feeding times, had little contact with its breeders. When Tessa was ten weeks old, she was purchased by Peter and Janet Taylor to act as a guard and as company for their young sons. The Taylors had never owned a dog before but had always liked the appearance of the GSD. They were drawn to the smallest puppy, who hovered shyly in the furthest corner of the kennel.

At her post-purchase veterinary examination, Tessa was very nervous, but responded well to gentle encouragement and titbits. Tessa's first experience of other dogs outside her own litter came when she was fourteen weeks old. A family friend brought his eleven-month-old GSD dog puppy to 'play' with her. The dog puppy was large and very boisterous, and pestered Tessa unmercifully, despite all her efforts to submit to him. Peter, mistaking Tessa's actions for play, let the puppies get on with their 'game'. At last, in desperation, Tessa flung herself at her tormentor, snapping and growling. The larger pup backed off in surprise, and Tessa's owner rushed to comfort her, cuddling her and using words of praise. Tessa had taken the first step along the path that, due to her owners' inexperience, her own inherently nervous temperament, the lack of early socialization and the lack of any understanding of even the rudiments of canine behaviour on the part of the trainer to whom the Taylors later took her, would ultimately lead to disaster. In a few seconds she had learned that a display of aggression not only discouraged a potential aggressor, but also gained the instant attention and praise of her owners.

Over the next couple of months, Tessa made every effort to avoid other dogs when out on walks. Worried about this obvious nervousness, Janet contacted the local dog club trainer, who suggested

1

that she take Tessa to the training sessions. On the first evening, Tessa jumped happily out of the car and followed Janet to the hall where the class was in progress, and found herself in an echoing room full of barking dogs. She slipped her lead, ran back to the car and cowered under it, refusing to be coaxed out. Janet managed to pull her out and with the help of the trainer dragged her back into the hall, where she spent the entire evening shivering under a chair. If it had been recognized at this point that Tessa needed to be gradually introduced to non-threatening dogs in a calm and friendly environment it is likely that her story would have had a very different ending.

Tessa was dragged into the hall again the following week, where she immediately took up position under Janet's chair. The trainer suggested that she take Tessa to the middle of the room, so that other dogs could be introduced to her. As the first dog approached, Tessa tried desperately to escape but couldn't, as she was restrained by her lead. She had learned already that in such circumstances submission was ineffective, so she flung herself at the advancing threat, growling ferociously. The dog's startled owner quickly pulled him back, while Janet petted Tessa, trying to calm her. Tessa's aggression had again been doubly rewarded. She remained very agitated, so Janet decided to give up for the night and have one final attempt the next week.

As soon as Tessa was dragged in through the door at the next class, her demeanour changed. She lunged forward to the end of her lead, intent on repelling any possible threat before it could approach her. Janet could not restrain her. The trainer ran up and grabbed at her lead. Tessa tried to bite him. Unfortunately, having no knowledge of the different causes of aggression or of the appropriate methods of treatment, he decided to teach her a lesson using the age-old methods. Tessa learned her lesson well, but it was not what the trainer had intended to teach her. She learned that strangers could be just as threatening and unpredictable as dogs – and she already knew how to deal with them!

During the following week, Tessa began to show signs of fear whenever she was approached by anyone she did not know. One evening, being busy herself, Janet asked her elder son to give the dog a quick run. A neighbour approached, hand outstretched. Tessa bit him hard and he ended up in hospital. The Taylors took Tessa straight to the vet and asked for her to be put to sleep. They were at once seen by a behaviour therapist who explained the reasons for Tessa's actions and assured them that with time and patience her fearfulness could be overcome, but the Taylors decided they had neither the time nor the commitment. Their son had been badly shaken by the episode, and they were worried that she might pose a threat to the children.

Tessa was put to sleep that same evening. She was just seven months old. It was a totally avoidable tragedy.

Editor's Comments
John Fisher

The case history you have just read was taken from Ruth Williams' own files. Ruth was the veterinary surgeon who offered behavioural therapy for Tessa and who euthanased her when the owners decided that the risk was too great. From her birth to her death, everything that could go wrong for Tessa, did go wrong, and the case highlights the need for a greater understanding of all aspects of our animal companions' behaviour. In this book are the answers to how this tragedy could have been avoided.

I am extremely honoured to have been asked to edit *The Behaviour of Dogs and Cats*, and at times my editor's pencil has trembled at the responsibility of presenting the work of so many knowledgeable authors.

The book starts by looking at why there is a need for behaviour counsellors, and what the future of behaviour therapy should be. It explains dogs' emotional development and how we can ensure that we give them the best start in life, then examines a variety of behaviour problems and how, with a greater understanding, prevention is always better than a cure. It also looks at the other end of the lead – at us – and shows that, providing our relationship is right, there are kinder and simpler ways of training dogs than the ones which have been on offer in the past. There follows an in-depth look at our other friend, the cat, and how we can overcome some of the problems which they can exhibit. The final chapter reveals that which we try to keep hidden – our emotions following the death of a loved pet.

I have certainly learned a lot through reading and discussing the various chapters with their respective authors. I am certain that what they have to say will make a major contribution towards improving the relationship between us and the animals we share our homes with. Tessa and her like desperately need our understanding.

1 A Vet's View of Behaviour Therapy

Sarah Heath

The concept of preventative medicine is very fashionable. We are showered with information about good diet, regular exercise and preventing heart disease from all sorts of sources, and are encouraged to see our doctors for regular check-ups and not just when we are ill. The veterinary profession also advocates preventative medicine. For instance, routine blood sampling for early detection of disease is now commonplace and preventative care for geriatric patients is offered in a large proportion of veterinary practices. Vaccination against the major canine diseases is now widely accepted and the annual booster appointments provide an ideal opportunity for a general check-up. The veterinary profession has wholeheartedly adopted the idea that prevention is better than cure, with inestimable benefit to the animals in their care. Nevertheless hundreds of dogs under the age of two years are euthanased every year because of some unacceptable component of their behaviour, and hundreds more are rehomed or even just abandoned. In the majority of these cases the problems could easily have been prevented if the owners had been educated in canine behaviour, which is why vets need to give more importance to this area within the field of preventative medicine.

No one today would dispute that health in a human context relates to both the physical and the mental components; equally, animal health should not be limited to the physical. Every veterinary surgeon admitted into the Royal College of Veterinary Surgeons makes a vow that 'my best endeavour will be to ensure the welfare of animals committed to my care'. It is my belief that the welfare of these animals includes not only their physical but also their mental well-being. Just as with human health, the symptoms of physical and psychological ailments in animals are not mutually exclusive and it is essential that full medical examinations are carried out first to ascertain whether any change in behaviour patterns is the result of some underlying medical complaint. For these reasons vets need to be involved in behavioural therapy.

Obviously it is not possible for every veterinary surgeon in the country to have a specific interest in behaviour or every veterinary practice to offer behavioural consultations at their premises, but there does need to be an increasing awareness of just how relevant behaviour is to the whole concept of animal health. We have come a long way from being the old-fashioned veterinary surgeons whose only involvement with small animals was merely to inject and dispense tablets. The waiting rooms of most small animal veterinary practices now reflect our involvement in the idea of responsible pet ownership. A wide range of pet-related merchandise is displayed, including chew toys, combs, leads and diets. There are also notice boards and leaflets displayed in many veterinary waiting rooms. Indeed the role of the veterinary surgeon in society is changing rapidly with more and more members of the profession becoming exclusively involved in companion animal work. The truly mixed practice is becoming a rarity and specialization is being promoted as the way to proceed. Pet owners are expecting and receiving medical and surgical expertise comparable with that provided by the medical profession. Despite these changes it is only recently that the profession has started seriously to consider the relationship between man and his non-human companions. On numerous occasions every day we hear clients describe their beloved pet as a dear friend and no one can deny that pet owners see their relationship as one of mutual trust and affection. On the agricultural scene the keeping of animals is justified on economic terms and yet despite the fact that man is becoming more and more obsessed with an input and output philosophy we continue to keep dogs and cats in our homes with no obvious role other than companionship.

It is my belief that the relationship between the veterinary surgeon and his client should start before the puppy or kitten is acquired. The vet and nurse are ideally placed to offer advice to prospective owners about which breeds would be suitable for their lifestyle, and also to give advice about what the commitment of pet ownership really involves. Some want a dog for a pet while others want protection for their homes, or a dog that can be used in field trials or for agility competition. The expectations of someone who lives in an urban block of flats are very different from someone who lives out in the wilds, and obviously not every breed of dog is suited to every owner, wherever they live. The reasons for wanting to own a cat may not be so diverse, and certainly there are not many important differences between one type of domestic shorthair and another. However, there are also pedigree cats and there are most certainly differences between breeds which prospective owners need to consider. The long coat of the Colourpoint or Birman may be an attractive feature but not many owners will be aware of the significant time commitment needed to groom these cats and keep them in good condition. There are temperament differences to consider, as, say, between the Siamese and the Maine Coon. Some veterinary practices stock copies of books giving details of different cat and dog breeds with their relevant characteristics, and these can be very useful when discussing breeds with prospective owners.

Once the new family member has been acquired the owner needs to be encouraged to approach either the vet or the nursing staff with any problems that they might have, including those of a non-medical nature.

The first encounter many owners have with a veterinary surgery is at the time of the first vaccination. This is one of the most controversial aspects of preventative medicine in the veterinary world for, although everyone is agreed that vaccination is vitally important, there has been and continues to be much discussion regarding the optimum time for the inoculations. The early life of a puppy or kitten is crucial to its behavioural development but one of the obstacles that people experience in trying to follow advice about the critical socializing and habituation of their pet from an early age is the time delay while they wait for completion of the vaccination programme.

Vaccination against the major canine and feline diseases is vitally important but it should not and indeed it need not jeopardize successful socialization, provided sensible precautions are taken. Specific advice about adequate socialization can be found in David Appleby's chapter in this book.

In the case of dog owners especially, the first and second vaccination appointments provide the ideal opportunity for veterinary surgeons to talk about socialization and habituation and to emphasize their importance, and to put the prevention of diseases by vaccination in context with the prevention of behavioural problems. Obviously these appointments are also important for discussion of such health topics as worming, diet and dental care, which means time has to be rationed as there is so much to be covered. For this reason some practices make the first and second vaccination appointments longer than the average consultation, but where this is not possible literature can be used to fill in brief outlines of particular points. Most veterinary practices today give some sort of vaccination package to the owners of new puppies and short information sheets about all sorts of aspects of the dog's health, including behaviour, can be included. David Appleby has produced two excellent booklets called 'How to Have a Happy Puppy' and 'The Good Behaviour Guide' which are now stocked by numerous veterinary practices. These booklets cover the topics of socialization and habituation and also the prevention of specific behaviour problems. They contain information not only for the new owner but also for breeders, who have a vital role to play in the behavioural development of the puppies that they rear.

It is very important to train puppies, yet many owners today still believe that they should not start training until their puppy is six months old. Indeed only recently a twelve-month-old German Shepherd Dog with a minor skin condition was brought to me and during the consultation the owner asked me for the telephone number of a local training club. She said, 'We want to take him to the dog club now 'cos he's twelve months old and that is the right age to start training, isn't it?' I ought to mention that this dog had been so terrified of me examining him that he had repeatedly tried to bite both myself and the owner and a muzzle had to be used. When I told the owner that this was most certainly not the case and that training should start as early as possible she became most indignant and told me that she had received the advice from a knowledgeable dog trainer. It not only saddened me to hear that such advice was still being given by someone who should know and understand more about dogs but I also felt disappointed that this client's expectation of a veterinary surgeon did not allow for anything other than a clinical understanding of our canine friends! I suspect that although as a profession we spend a considerable amount of time in the company of animals we all too often do not take time to observe them and so deepen our understanding of them. Consequently people have looked elsewhere for advice and have pigeon-holed veterinary surgeons as only being concerned with animals' ailments. It is my belief that being a vet should involve far more commitment to our companions than that and that veterinary practices should be a source of advice and information on all aspects of pet care.

Vaccination appointments are the ideal time to draw owners' attention to the idea of attending puppy socialization classes or puppy playgroups when the vaccination course is complete. The idea was first introduced by Dr Ian Dunbar and groups are now being run throughout the country. Veterinary nurses are ideally placed to run these groups and many have found it a very rewarding addition to their role within the practice.

Because the owners often find the concept of playgroups, and indeed any training for very young puppies, strange at first, it is important for the vet and the nurse to take time to explain why these classes are so beneficial. I can certainly vouch for their effectiveness. I often observe a phenomenal difference in puppies that attend when they next come to the surgery.

The groups are aimed at the pet dog and are designed for puppies up to the age of sixteen to eighteen weeks. The age limit does not need to be rigid and some individual variation must be allowed for but generally puppies over this age have passed the optimum time for learning such social skills as bite inhibition and may indeed pass on bad habits to others. Some practices may have suitable rooms in which to hold the classes while others may need to hire a local village hall, for example.

Puppy playgroups are not a free for all where a number of puppies are allowed to run riot. If this were the case the pups would merely learn to play rough games which their owners would be unable to control. The idea is for the sessions to consist of controlled play sessions interspersed with training so that the puppies become socially acceptable dogs which are a pleasure to own. Groups of between four and eight puppies seem to be ideal in order to allow for adequate socialization and a degree of individual tuition, and a session of between an hour and an hour and a half gives the puppies and their owners the time to settle into the groups but avoids problems of the puppies becoming over-tired and disruptive. There needs to be a balance within the sessions between educating the owners and educating the dogs since both need to be educated if the group is to be successful. Good communication between the instructor and the people attending the class is necessary in order for the owners to understand the way in which puppies learn.

Socialization with people, dogs and possibly other animals is vital at this early age and so it is helpful for as many members of the family to attend the groups as possible so that the puppies present can meet men, women and children. Owners need to be shown that this process must be continued at home between sessions. Teaching and training are the two other components of the classes with the importance of teaching a puppy not to jump up and training it to sit on command being equally emphasized. Anyone who is keen to learn more about the practical considerations when starting up a group should read a booklet entitled 'Running Puppy Classes' written by Erica Peachey of the Association of Pet Behaviour Counsellors. Puppy playgroups are an ideal first step toward responsible pet ownership and if it is not possible for the practice to run classes then the members of the practice can at least find out where the nearest classes are being held and encourage their clients to attend.

Puppy parties at the surgery, another idea that is gaining ground, are an ideal opportunity for owners of new puppies to get to know their local practice and learn about the various aspects of puppy care while helping their puppies in the important steps of socialization. The parties are open to puppies between their first and second vaccinations, which will ideally be between the ages of nine and twelve weeks depending on the vaccination policy of the practice. Each puppy only attends one party and therefore one evening a month is all that is required of the vet! The parties are held at the practice and last for one to one and a half hours. The practice provides light refreshments for the owners as well as titbits and toys for the puppies, and an informal feel to the evening helps owners to relax and ask questions. The actual format will obviously vary from practice to practice but generally an introduction of the people present is followed by an introduction to the practice and the services it offers. The puppies are then gradually allowed off their leads so that they can socialize and get to know each other. All of the children present will then be encouraged to handle the puppies, it being as important to teach children how to approach puppies and handle them correctly as it is to teach puppies to accept children! The adults are then shown how to handle their own dogs correctly and are given the opportunity to handle each other's by playing 'pass the puppy'.

The next part of the evening will include short talks by the nurse or vet on such health topics as vaccination and the need for annual boosters, regular worming and ectoparasite control. This is followed by talks on behaviour, including the importance of socialization, the control of chewing and mouthing and the basis of the relationship between dogs and their owners. Training and teaching will also be discussed, and an introduction given to basic obedience, including sit, stay, come, etc., house-training and walking on the lead. Veterinary nurses with a specific interest in these areas are well able to give these talks or alternatively a local behavioural counsellor can be invited to come along and speak.

After this more formal teaching session the puppies are taken into the consulting room one at a time and put on to the table where the vet or nurse will fuss and feed it with a small titbit in order to give pleasant associations with the room. This is followed by a play session where the owners are encouraged to play the right sort of games with their puppies. An informal question and answer time ends the evening with refreshments then being served to encourage people to mix and chat. Practices which offer these parties to their clients find that the puppies which attend are generally easier to handle on future visits to the surgery – and certainly the owners are more relaxed and at ease with the practice.

It is my hope that one day every practice in this country will provide such services as I have described and eliminate many of the problems we face daily. Until this comes about, there is a great deal that the vet in practice can do for owners of dogs which have developed behavioural problems. First, we can listen while remembering that nearly all behavioural problems are a manifestation of normal behaviour in inappropriate circumstances. We are often approached by owners who can no longer cope with the strained relationship they have with their pet and who ask us to suggest methods of rehoming. In the extreme cases the veterinary surgeon is quite literally the end of the line and euthanasia is performed. In the area of canine behavioural problems there is a fundamental misunderstanding between our two species. By far the majority of dog owners think of their canine companions as 'one of the family' and talk to them regularly. We talk in terms of our friendship with them and extol them for their faithfulness. In fact we subconsciously turn our dogs into little people. Even farmers do this, who consider that their dogs must work for their living. They often have one of the closest pet–owner relationships and almost all of them talk to their dogs in more than simple command language. There is no reason to criticize this anthropomorphic view of man's best friend but it can result in us viewing their behaviour in human terms and often losing the ability to see them as the domesticated wolves they really are.

Fortunately for man by far the majority of our canine companions do fit into their man-made roles very well, but for some the 'language' barrier between man and dog is too great and communication breaks down. Often the owner will then approach the vet with concerns about their pet's behaviour and as a caring profession we need to take these problems seriously and endeavour to find practical solutions. Certainly some of the anxieties clients have will turn out to be related to training problems and in these cases it is sensible to refer the client to a local reputable dog trainer. Here we must be careful not to give the phone numbers of training clubs we have never visited and about whose training techniques we know nothing. Many people will assume that the training club is in some way approved by the practice, especially new dog owners who have no experience against which they can judge the trainer. Inappropriate training techniques can do immeasurable damage to a dog and therefore practices owe it to their clients to suggest only training classes that they know something about.

Where a presented problem is clearly a behavioural one, we need to take a different course. Vets in general practice see examples of behaviour that they would classify as a problem everyday. The range of muzzles stocked by most practices is evidence of this. Whether we are dealing with a fearful dog in the consulting room or an overtly aggressive one on a house call the use of a muzzle can sometimes be the only way to ensure that the job gets done safely. All too often relatively simple examinations need to be carried out under sedation because the dog's behaviour makes examination in the consulting room impossible. However in many of these cases the owners are perfectly happy with their pets and we need to remember that the definition of a behaviour problem is any behaviour identified by the owner as being detrimental. The treatment of behavioural problems depends on commitment on the part of the owners since all behavioural therapy involves relatively long-term treatment and consistency of application is vital to its success.

Behavioural problems can of course have a medical cause, examples of which include circling and head pressing due to increased toxin levels in the blood associated with hepatic encephalopathy, or the less dramatic but nonetheless distressing behavioural changes associated with false pregnancy in the bitch. Urinating in the house may result from a medical problem involving the urinary tract rather than from a behavioural problem of urine marking. It is important to make a full medical examination in cases where owners report what they consider to be a behavioural problem. In many cases there is a need for a combined approach to treatment using both medical and behavioural therapy and the hormonally driven behaviours, the mounting of table legs, for example, are good instances of this.

There is a range of drugs available which can be used in the treatment of certain behavioural conditions. However we must guard against the tendency to reach for the drugs in every situation and must not be misled into believing that any one drug is a cure-all for behavioural problems. Pharmacological preparations must be used in conjunction with behavioural modification techniques if we are to avoid the situation where the problem returns when the course of tablets comes to an end.

Most vets just do not have the time to spend with individual clients in order to establish the causes of problems and institute suitable programmes. It is therefore essential that they have available to them professional people to whom they can refer their clients in the sure knowledge that they will receive not only good sound advice but also the necessary counselling. The Association of Pet Behaviour Counsellors (APBC), a nationwide and indeed international network of experienced professional behaviourists available exclusively on referral from vets, provides such a service. For many the mention of animal behaviourists conjures up a mental picture of a dog or cat lying on a couch while a bespectacled psychiatrist asks deep and meaningful questions. However the reality is that the workload of behaviourists in the companion animal field is steadily increasing. They are not wacky psychologists but people with an understanding of animal psychology that translates into practical techniques. They assess the problem carefully and then offer advice to the client regarding treatment regime, time scale and future options. Members of the APBC keep in close contact with the referring vet and by providing telephone follow-up and aftercare support they help to motivate the owners to continue the therapy.

I am proud that the veterinary profession is held in such high regard by the public and honoured that they trust us with the well-being of their beloved pets. In return for that trust we owe it to owners and pets alike to take behavioural problems seriously. If veterinary surgeons and behaviourists work closely together we can aim for that ideal pet–owner relationship to the mutual benefit of man and animal.

2 The Development of Social Behaviour

Roger Abrantes

What makes a social animal special is its ability to control dominance and submission. Fear and anger, the fight for survival, the ability to find food or sexual partners, alliances with other species and so forth are traits we find in many different animals belonging to many different species. Some of them, however, will have a greater ability in coping with individuals of the same species. They are skilled in the use of dominance and submission, they are social animals. Among these animals we find wolves, geese, chimpanzees, humans and their best friends, dogs.

In the beginning . . .

A newborn puppy seems utterly helpless, but it is not so. It cannot see and yet it is able to find its mother's nipples and begin sucking. It does not hear very well and its sense of smell is not yet well developed, but it will whine and wait if it is separated from its siblings, or when it is cold and lonely, until its mother retrieves it back to safety. At this time there are no signs of the great mechanisms of dominance and submission which will play such a large part in its life. At this early age all the puppy's energy, in common with all other newborn creatures, is concentrated on survival.

Warmth and food are its two most pressing necessities and they are given freely by the bitch. There is no need for aggression, dominance or submission. Fear does not yet work to full force and is only elicited by what is physically unpleasant such as cold, loud sounds, *unpleasant* or *painful* experiences.

Fear

What we call *fear* is a stress reaction to anything that seems dangerous. It elicits a series of physiological and anatomical processes in the dog that all aim at the best possible solution for its survival. For the puppy in fear there are several available alternatives open to it, such as to

Figure 1 *The newborn puppy seems helpless and yet it is able to perform all the activities necessary for it to survive. The newborn puppy is a perfect survival machine. It needs food, warmth and contact and it will draw attention to itself if one of these is missing. An interesting rearing behaviour concerns the mother's overturning of the puppy to lick its belly. At first the puppy finds this disagreeable but then learns that it is pleasant, not least because it stimulates digestion. Perhaps we have here the foundations of passive submission where the dog lies on its back.*

retreat, to whine, or to lay down paralysed and yelling. Generally it may be said that fear leads to *flight* or *passivity*.

Fear is of the utmost importance for the survival of the individual and is probably inborn. Without it no individual survives long enough to be able to reproduce itself and perhaps pass this trait to its offspring. Immediately after birth, fear is probably only elicited by unpleasant physical stimuli, but as the puppy becomes keener to recognize factors or situations which might be unpleasant or dangerous it will be increasingly aroused by other stimuli. Clearly there is some mechanism at work that recognizes the unpleasant as a danger for the welfare of the organism.

Fear calls into operation flight or passivity mechanisms that must be efficient and reflexive if they are to save the life of an individual who is one of the species that are potential prey. For instance, passivity must be convincing enough to eliminate danger, particularly if one is a little bird a few yards away from a predator or a rabbit 20 yards from the fox. Here total passivity is the best possible behaviour, for flight is not likely to be successful. And it is vital to show total passivity in certain confrontations with much bigger and stronger members of one's species in order to pacify them.

The young puppy must have the proper flight and passivity patterns in order to survive. At this very tender age the puppy has not the slightest possibility of fighting back. Of course, as it develops physically other alternatives arise and fighting back becomes a viable option.

Aggression

Between four and five weeks puppies begin to show the first signs of aggression. At this age they are more aware of the world around them, more conscious of themselves.

The first confrontations seem to arise without cause as there is no obvious reason for them. There is food and care enough for everybody and they are not yet interested in possessing things, which is a usual cause of dispute. Perhaps it is because pups *are* together and at this phase of their development are not yet social animals. Dog pups and wolf cubs do not start out social animals. They *become* social.

All newborn are selfish, almost by definition. Later they may lose some of this egoism and learn to be social. In fact what happens is that selfishness takes some more sophisticated form – not a pursuit of an immediate advantage, but a long-term benefit. This is what being social means, and this has to be learned, although most definitely there is a genetic disposition to it, which may work like the genetic disposition for the development of certain coat colours, not at birth, but developing later.

Soon puppies want the odd plaything or bone for themselves. Confrontations are in the beginning confined to siblings. They are not

Figure 2 *Fear is expressed through many behaviour patterns that are always combined with submission. When the dog lays down quietly it is showing passive submission. When it nuzzles its opponent with ears laid back, lips pulled down and back and small eyes it is showing active submission. Flight is always an expression of fear and submission.*

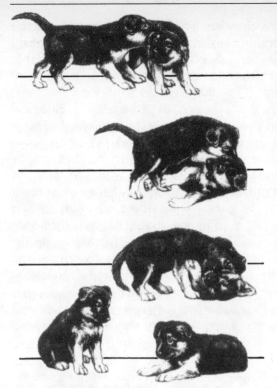

Figure 3 The first encounters with aggression are through interactions with siblings. Controversies bring pain which elicits fear until the situation is under control. Because the sibling will bite back, the puppy soon learns to stop biting the sibling to attain what it wants. The puppies are, unknowingly as yet, using dominance and submission in their encounters.

dangerous and the puppies suffer no harm. Loss or capture are taken with equal serenity. They are learning to interact even if they are not aware of it. This is the plan of the special genetic programming that drives them.

Aggressiveness is elicited when there is a need to resolve or promote conflicts between members of the same species. The more your opponent looks like you, the more reactive you will be. Sometimes you compete with individuals of other species, over the odd carcase, for instance, but the competition will be fiercest with individuals of your own species, and worst of all if they are of the same sex as you. Individuals of the same species compete over food, sexual partners and territory, undoubtedly the three most important things for most organisms. Like fear, aggression is a stress reaction to a potentially endangering situation. An opponent is always a danger unless you control him or her totally.

At a certain time in their development, supposing they live a normal pack life with mother, father and others, puppies meet three different circumstances which will change their lives forever. It is a hard time, it is a learning time. The three circumstances are competition with siblings, weaning and their father.

Puppies meet the urge for aggression for the first time in disputes over trivial items, such as the odd bone. They fight with each other and, without knowing it, receive their first lessons in how to become a social animal. Soon they realize they cannot handle siblings by means of aggression and fear and then other strategies develop.

At the same time their mother denies them the best source of food they will ever have, suckling. The time to obtain goods for free is over. Every time they see their mother, they try to get to her to suckle, and sometimes, if they join forces and surprise her, they succeed. This is perhaps the first sign of co-operation among puppies. Alone they have no chance. The mother seizes them by the nose, pins them to the ground, growls at them and they run away whining and howling.

Also at roughly the same time they meet their father or some other grown-up male member of the pack for the first time. It is the first occasion when they have met a complete stranger. Until then they have only been with their mother and siblings. In the beginning the father is tolerant and complacent, but soon there are controversies, sometimes over trivial matters. He is big and strong. He is tender but he is also a bit awe-inspiring. The puppies fear him in the beginning. They run away whining as they did when they had controversies with their mother during weaning. Then suddenly they seem to change their tactics.

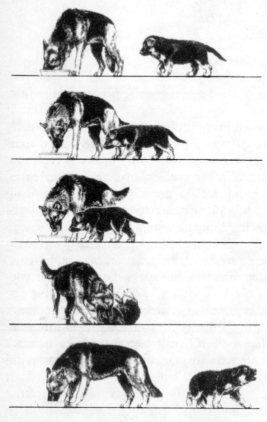

Figure 4 *Around the time of weaning the puppy experiences aggression from its mother for the first time. It may be in a dispute over food. Usually the mother snarls and growls at the puppy, which does not react. The mother then attacks the puppy by grasping it by the muzzle and pinning it to the ground while it whines and yells.*

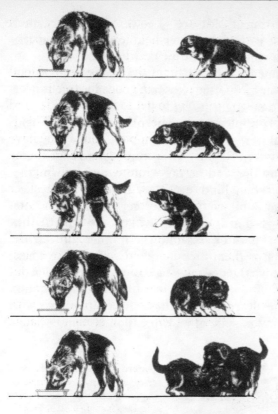

Figure 5 *Soon the puppy learns how to deal with its mother's aggression. In the beginning the puppy is undoubtedly fearful, but as soon as it is able to cope with the situation it does not become fearful. Submission is enough – a series of behaviour patterns related to neonatal behaviour which then brought the puppy pleasant results such as milk and contact. The paw waving is related to the stimulation of the mother's milk production and the hind leg twist is related to the mother's licking of its belly to stimulate its digestion.*

Submission works because it allows the pup to retreat without harm and perhaps even engage in sibling biting activity.

Submission

When two puppies of about six weeks play they bite each other, and if suddenly one of them gets a good grip on the other's ear, it bites really hard and the poor brother or sister understandably howls. Now roles reverse. If, even with its ear firmly grasped, the puppy succeeds in getting its teeth in the tender skin of its sibling's belly, much whining and yelling follows. They both let go and look puzzled. Having learned their lesson they will act a little differently next time. One will grasp the other's ear only until it begins making a noise, then it lets go because it knows that if it doesn't the other will hurt its belly. This is what is called learning by *trial and error*.

Another situation. The mother is chewing a bone. The puppy comes closer. She growls at it in warning. The puppy ignores the warning and continues towards the bone. When it is 10 inches from the bone the mother seizes the puppy by the muzzle and pins it to the ground in a flash. The puppy whines and runs away. Ten minutes later the puppy again approaches. This time the mother growls and the puppy stops and begins displaying the whole range of what we call pacifying behaviour. It licks the air with its tongue, twists with the hind leg and waves one front paw. The puppy is displaying the same movements that used to bring it pleasure. The licking is associated with suckling, the twisting with the mother licking its belly and the front paw waving with the stimulation of milk production. And it all works. The mother continues chewing her bone and the puppy may safely run away. Another very important lesson has been learned: how to deal with aggressive and stronger individuals.

A third situation. The puppies come out of the den. We still suppose they live a natural life, or that they are wolf cubs. They see their father 10–15 yards away. Rapidly they run towards him and nuzzle him, try to lick his lips and at the same time perform the characteristic twisting with their hind leg. They lie down, belly up, quickly stand up again and follow the father repeating the whole procedure. The father growls a little at them. After some time he will seize them by the muzzle, more or less like the mother did during weaning. As soon as he grasps them they lie down voluntarily. Then they seem to be satisfied, leave the male in peace and quickly engage in playful activity.

So what have the puppies learned from these three situations?

The puppy bites its sibling in spite of its expression of pain because

Figure 6 *Paternal behaviour does not differ much from maternal behaviour with some obvious differences related to anatomical and physiological factors. The dog or wolf father is normally tolerant and friendly towards the puppies. Sometimes they pursue him and seem to insist on being seized by the muzzle and pinned to the ground. The whole thing is actually much less dramatic than it looks. The puppy does not need to be pinned to the ground as it promptly lies down. The whole ceremony is a test of the father. The puppy needs to feel safe. It needs to know that the father, big, strong and impressive as he is, is still in full control of itself and will not harm it.*

Figure 7 *Aggressiveness is elicited when there is competition primarily between individuals of the same species. It can also occur among individuals of different species when they compete for the same food item. Aggressive behaviour patterns can be seen in conjunction with dominance or submission. When they occur together with submission it is always because earlier active or passive submission has not been accepted and flight is impossible. The animal has no other option but to fight back, turning fear into aggression while keeping the submissive attitude.*

it does not know that it is inflicting pain on the other. It learns it through its own feeling of pain when its sibling's teeth grab its belly. Whether or not it realizes what pain is for the other does not matter. What matters is that it learns to act in accordance with the other pup's message of pain, its whining, and it stops biting.

The dog or wolf mother is not teaching the puppy a lesson in aggression. On the contrary she is teaching it a lesson in how to avoid aggression, to compromise, i.e. how to survive when you get your way and when you don't. This is perhaps the most important lesson a youngster may learn for the ability to compromise is the essence of the social animal.

The dog or wolf father is not teaching the puppies to fear him. If he elicited fear, the puppies would run away instead of towards him. So why do they keep following him and bothering him? They seem to want something from him and it is that they want him to show them that he accepts them and will not harm them. It is a test of self-control. 'Show us that you can grab us with your formidable jaws without harming us,' they demand. 'Show us your self-control. Show us that we can feel safe with you.' As their father he can fail the test in two ways, either by ignoring them, which shows he is not their true guardian, or by hurting them, which shows he is a bad guardian they cannot trust.

All these lessons aim at one thing, to teach the puppy how to compromise, which means how to use the mechanisms of submission and dominance to perfection so that it does not need to revert to the use of fear and aggression.

Dominance

From the very first time the puppies learn to submit themselves to either mother or father, they are simultaneously learning dominance. In the beginning they will use it only in confrontations with siblings. After the first episode of the ear and belly biting, the puppies are bound to repeat the situation, to improve and perfect the technique. This means not biting too hard, but making the other yell as soon as possible. The puppies are trying to perfect their use of submission and dominance. In the beginning it is easier to use submission for they are

Figure 8 *This shows the evolution of a behaviour pattern. Social behaviour is not inborn, it is developed from patterns which had a function at an earlier age. By becoming social behaviours they often lose their original function and obtain a different meaning. When puppies lick just after birth it is in connection with suckling. Later they lick adults' lips to make them regurgitate food. Still later this licking becomes a pacifying gesture. It is not food they want any longer, but acceptance and friendliness.*

Figure 9 *Fox, jackal and wolf – three different animals, three different strategies from nature. The fox hunts alone and does not need many facial expressions for communication. The jackal lives in small families of an adult male, an adult female, a juvenile and puppies and needs a greater range of expressions than the fox. The wolf has a very developed range of communication patterns. The wolf is a highly social animal and in a wolf pack there are many animals and several of the same age group and of the same sex. Among these there is a high rate of confrontations that must be resolved by other means than aggression and fighting. Repeatedly aggressive action exhausts the individual and endangers other activities such as mating. In the end the fittest will survive. The fittest is the one able to reduce the amount of antagonistic activity to the advantage of reproductory and nursing activities. This is exactly what makes the* social gene *survive from one generation to the other at the cost of the* fighter gene.

used to that in the confrontations they have with adults, and need to master it. Soon they also need to control dominance because they begin to compete seriously with their siblings over food and other items.

At last they become masters in the use of these mechanisms and do not waste time or energy in unnecessary displays. They live in a world where energy is needed for survival and where waste of it is heavily penalized. Only the fittest will survive long enough to give their *fittest* genes to their offspring. Among these social animals, the fittest are undoubtedly the best in using the mechanisms of dominance and submission, and their offspring will be even better.

In the end there is only one purpose: to live as long as possible, as well as possible and preferably long enough to allow for reproduction of the genetic information. Beyond that there is no meaning. Some organisms have to share space and time with members of their own species and sometimes with strangers. As long as they all achieve their own goals, they can share the same environment, and they will develop behaviour patterns accordingly. So simple is nature's strategy – the strategy of life!

References

R. Abrantes, *Hundesprog*, Borgen Forlag, Valby, Denmark, 1985.

F. W. Christiansen and B. Rothausen, *Behaviour Patterns inside and around the Den of a Captive Wolf Pack*, unpublished scientific paper, 1983.

C. Darwin, *The Expression of Emotions in Man and Animal*, The University of Chicago Press, London, 1872.

K. Lorenz, *Das so gennannte böse*, Dr. G. Borotha-Schoder Verlag, Vienna, 1963.

—*The Foundations of Ethology*, Springer-Verlag, New York and Vienna, 1981.

D. Meach, *The Wolf*, The Natural History Press, New York, 1970.

E. Zimen, *Der Wolf: Mythos und verhalten*, Meyster Verlag GmbH, Vienna and Munich, 1978.

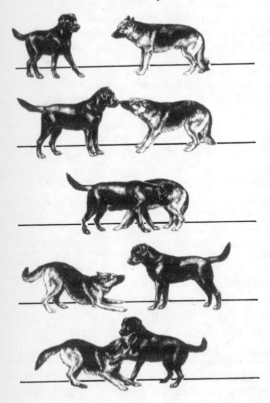

Figure 10 *When two dogs meet they perform the whole range of behaviour patterns associated with greeting. Greeting is a very important ceremony between animals that are both social and aggressive. Greeting rituals develop in order to control aggressiveness. In the greeting ceremony both sides ensure each other of their non-violent purposes and at the same time signal how far they will go in dominant/submissive behaviour. A clear line from the beginning is the best insurance for future good relationships. Social animals are specialists in developing and performing greeting rituals.*

Figure 11 *Social behaviour explained by means of the concepts of fear, aggressiveness, submission and dominance.*

3.2 shows a neutral expression. From 3 towards 1 dominance increases and from 3 towards 6 submission increases. In each sequence, from .2 towards .0 aggressiveness increases and from .2 towards .4 fear increases.

All the 3.2 illustrations show dominance and submission attitudes without aggressiveness or fear. 1.2 is the leader of the pack and 5.2 illustrates cubs.

3.3 shows the greeting ceremony, which always includes elements of submission. To be friendly is, to a certain degree, to submit one's needs and wishes to the other's.

All drawings are by Alice Rasmussen from the book *Hundesprog* by Roger Abrantes and are reproduced by courtesy of Borgen Publishers, Valby, Denmark.

3 Socialization and Habituation

David Appleby

I'm sure most of us know the parable that compares the foolish man who built his house on sand with the wise man who built his house on rock and their subsequent failure and success. This story can be used as an analogy for the appropriate and inappropriate methods of raising puppies. The consequences of using inappropriate techniques can be very severe: the major cause of death in dogs under two years of age is euthanasia due to behavioural problems (McKeown).

One in five of the dogs that Dr Valerie O'Farrell studied while conducting research at Edinburgh University Veterinary School had a behavioural problem to a lesser or greater extent. A similar, but larger, American study fixed the figure at one in four. In one year I treated 773 dogs – 79 of them, that's 10 per cent, had problems of fearfulness towards people or the environment due to a lack of early socialization or habituation and a further 4.5 per cent were inept at relating to other dogs, again due to a lack of early socialization. It is pure conjecture, but if one in five of the country's 7 million plus dogs has some form of behavioural problem, and 14.5 per cent of those have problems due to a lack of socialization or habituation, then there could be as many as 200,000 dogs with problems that are a direct result of their early experiences, or, more to the point, lack of early experience. I actually believe the problem is immeasurably greater than these figures suggest. Many dogs show a weakness of temperament or inability to cope when faced with a particular situation, without their behaviour becoming problematical enough for the owners to seek help from a behavioural counsellor, so are those that do just the tip of the iceberg?

What are socialization and habituation?

Socialization can be described as the process whereby an animal learns how to recognize and interact with the species with which it cohabits. In the wilds this is likely to be limited to the animal's own species, but for a dog in the domestic environment it includes other species such as man and cats. By learning how to interact with these the socialized dog develops communication skills which enable it to recognize, amongst other things, whether or not it is being threatened and how to respond.

Habituation can be described as the process whereby an animal becomes accustomed to a non-threatening environmental stimulus, such as a repeated noise, and learns to ignore it.

There is a critical period of development in which socialization and habituation must occur and be properly completed if the dog is not to grow up maladjusted. The degree of deprivation a dog suffers in respect of socialization and habituation will be reflected proportionately in the extent of maladjustment. Accordingly, a dog that has had no experience of a specific stimulus at the completion of the critical period will always be fearful of it; a dog that has had some exposure, but not sufficient, will be better adjusted, although not entirely sound; and a dog that has had adequate experience of the stimulus in the critical period will grow up to be 'bomb proof', as they say in horsey circles. A stimulus can be anything within the dog's world, a man, woman, or child, heavy goods lorry or a windblown black plastic bag rolling down the street.

The evidence which shows the crucial importance of systematically socializing and habituating puppies during the critical period has been around for a long time. However, those who advocate acting on the conclusions it points to are lone voices crying out in a wilderness of apathy and ignorance. Let us look at some of the research that has been carried out on this question.

Few people interested in animals can be unfamiliar with the imprinting experiments of Konrad Lorenz, who, from the 1930s onwards, was recording the fact that birds such as geese hand-reared from hatching became imprinted upon him and behaved towards him as to a parent of their own kind. In fact, Lorenz found that birds would imprint on virtually anything, even a flashing light, and treat it as a mother. Significantly, birds that had accepted Lorenz or a bird of another species as a surrogate parent would also recognize and accept other people or members of the adopter's species. Similar experiments have been carried out with other types of animals, all of which have confirmed the important role imprinting has on species recognition and subsequent social and even sexual orientation. There is a good story on record about a hand-reared bull moose that became amorous with his keeper rather than the female moose with whom he was supposed to be having an assignation.

But does the principle of imprinting on other species hold good in the domestic dog, you may ask? The answer is a resounding yes, although it takes longer than in some other species. The imprinting period for puppies is longer than that for, say, chicks simply because puppies are born blind and deaf and relatively immobile so are not fully able to start the process of species recognition at birth. However, an experiment on this question was conducted at Utrecht University where half of a litter of newborn puppies had no exposure to humans while the other half were exposed to a high level of human scent for just one minute, after which the complete litter was kept in isolation from human contact for several weeks. When they were reintroduced to human company, it was found that the puppies that had received the early exposure to the researcher's scent had a distinct preference for investigating people as opposed to investigating other environmental stimuli, whereas those puppies that had not had the early experience showed no preference. In 1965 Scott and Fuller identified the age of three weeks as the start of a puppy's critical period, in terms of social/environmental interaction and the commencement of their capacity to develop social relationships. Significantly, this is the point in time when the puppy becomes truly mobile and can hear. Indeed it coincides with increased electrical activity in the brain.

In 1971, Michael Fox, a behavioural researcher, found that three-week-old Chihuahua puppies fostered individually in litters of four-week-old kittens would by twelve weeks prefer the company of cats over the company of their litter mates that had not been fostered. Additionally, the foster mother's kittens were found to be able to relate to dogs whereas kittens from other litters who had not had a canine companion thrust upon them avoided contact with dogs. In the same year Michael Fox carried out a subtler but even more revealing experiment. Litters of puppies were split into three groups: one group of puppies were hand-reared from birth and received no canine contact; the second group were given an equal amount of canine and human contact; and the third group only experienced the company of other puppies and their dam. When these three groups of puppies were finally reintegrated the puppies who had only experienced the company of humans tended to select the company of other puppies who had only experienced human interaction. Similarly, those puppies who had been exposed to both human and canine company preferred the company of puppies of the same upbringing, as did the puppies only used to canine company. Perhaps the most significant tests of all are those similar to the ones carried out in 1961 by Freedman, King and Elliot and in 1965 by Scott and Fuller, which found that if puppies are kept in isolation from man and introduced at different ages their response to man deteriorates with age. The results show that if puppies are introduced to humans for the first time between three to five weeks of age they will approach confidently, but those that are introduced between five and seven weeks of age will show increasing amounts of apprehension. Those puppies whose first experience of man is at nine weeks old or later will be totally fearful.

In 1968 Scott concluded from his research into puppies kept in isolation from man until fourteen weeks 'by fourteen weeks fear and escape responses have become so strong that any puppy raised in these surroundings acts like a wild animal'. Freedman, King and Elliot also found that puppies exposed to human company at fourteen weeks for the first time never developed a positive approach, in other words they remained wild.

What does all this and other research tell us? First, even newborn puppies can achieve some level of recognition and imprinting to those around them by scent. (After all, they have to be able to recognize and locate the milk bar!) Second, that the critical period for a puppy's socialization starts at around three weeks old, that between five and nine weeks old a puppy's response to new stimuli is increasingly fearful and that after fourteen weeks there is no chance of the puppy recovering.

Why, one may ask, does a fearful response develop, even if puppies don't actually have an unpleasant or fear evoking experience in association with a stimuli? One would suppose that puppies would continue to be confident in the presence of all new stimuli if they have not had an unpleasant experience associated with novel stimuli at any time. The answer is that in their natural environment wild canids, specifically the wolf, to whom the domestic dog is related, have to be alert to danger, which means treating anything with which they are not already familiar as potentially hazardous. The older the individual is when it encounters a new experience the more extreme its negative reaction is likely to be. This means that a wolf cub, for example, has only a few weeks to develop positive associations with its own kind and immediate environment, after which it becomes increasingly cautious about things and situations it has not previously encountered. This saves it from blithely trotting up to something like a snake and investigating it. The problem the domestic dog has is that it needs to become familiar with an enormous number of stimuli in a very short time so as to be able to live in and cope with our rather artificial world.

So far, the research I have cited has been concerned with aspects of socialization, but what of habituation, i.e. environmental stimuli rather than social interaction? Experiments have been designed to reveal a puppy's critical habituation period: for example, puppies housed in conditions devoid of stimulation were placed in a test area with various articles for just half an hour at five, eight, twelve and sixteen weeks. These puppies were found to be increasingly keen to explore the items and to develop a preference for those that provided more complex stimuli. However, puppies who did not enter the test area until they were over eight weeks old tended to withdraw from rather than explore the items, and those who did not experience the test area until they were twelve or sixteen weeks old frequently

27

became paralysed with fear (Fox 1971). These results agree with those from socialization tests, reinforcing the theory that there is a critical period in which a puppy needs a stimulus-rich environment and social interaction. Experiments have also found that animals of various species who receive plenty of stimulation early in life subsequently have a good capacity for coping with stress and have a preference for exploring complex stimuli.

Conversely, experimenters have found that animals, including humans, who are deprived of stimulation in early life respond less favourably to stress and are not inclined to explore complex stimuli. This has to be significant for anyone who is interested in dog training as it is essential to the success of training that a dog is able to cope with stress and has a positive response to complex stimuli and situations. Stress inhibits learning, and training requires of the dog the capacity to process complex stimuli and associate the responses required from the handler/trainer.

What practical applications do we have that bear out the research? The answer is that there are accurate records for over 24,000 puppies over 25 years, all of which have been systematically socialized to man and habituated to the environment from six weeks of age, which form in effect a massive field study with quantifiable results. The work has been carried out by Guide Dogs for the Blind, who, until 1956, used to rely on the donation of adult dogs which they took on approval to maintain their training stock. The success rate of those dogs fluctuated between 9 and 11 per cent and it was recognized that this could be improved if the association could supervise the rearing of puppies. These were purchased and placed in private homes at between ten and twelve weeks old or even later. Things improved, but the results were not good enough. It was Derek Freeman, who subsequently received an MBE for his work, who pushed to have puppies placed in private homes at an earlier age to enable them to receive the maximum socialization and habituation during the critical development period. Derek had a strong belief in Scott and Fuller's work on the importance of early socialization and habituation in the production of dogs that were best able to survive and perform in the world at large.

Derek found that six weeks was the best time to place puppies in private homes; any later critically reduced the time left before the puppies reached twelve to fourteen weeks; but if puppies were removed from their dam and litter mates before six weeks they missed the opportunity to become properly socialized with their own kind, which then resulted in inept interactions with other dogs later in life. The training success rate soared because of this policy, which was carried out in conjunction with the management of the gene pool via the breeding scheme Derek also pioneered. Annual success rates in excess of 75 per cent became common. You might think that because this is a special scheme for dogs with a special function that it is not necessary to manage dogs who are to become household pets in the same way. In fact, what the scheme provides is adult dogs that are well socialized and habituated as well as of sound temperament. It is these dogs that *coincidentally* make the best material for training as guide dogs. As a result of the breeding scheme, Derek Freeman also proved, if proof was needed, that you cannot dismiss the importance of genetic predisposition, i.e. the basic material required for good temperament can be produced through good breeding. Conversely, a lack of habituation/socialization can ruin the chance of an individual developing a sound temperament, however good the genealogy.

Before I turn to other questions, there is one more line of research that must be mentioned because it introduces another parameter within which dog owners, breeders and trainers etc. are obliged to work if a puppy's potential is to be maximized. The research has revealed the fact that socialization and habituation can wear off. J.H. Woolpy's work with wolves in 1968 (Rab, Woolpy and Ginsberg) showed that adult captive wolves can be socialized with man with six months' careful handling. This was highly skilled work carried out under very artificial control conditions and remained specific to those conditions, and the team of skilled researchers involved reported that the experiment was very dangerous. The researchers found that if those wolves subsequently had less contact with them, their level of socialization did not regress, but wolf cubs that were socialized in the optimum period, i.e. up to twelve to fourteen weeks, lost their socializing capacity when interaction with the researchers was withdrawn. Michael Fox has stated that if well-socialized puppies are placed in a kennel environment between three and four months of age, and left there in virtual isolation until they are between six and eight months of age, they will be shy of strangers and even of their caretakers if they have not handled them much, whereas well-socialized puppies of six months old or more left in the same surroundings for two months do not lose the socialization level they have achieved. Woolpy, from his research findings, concluded that although there is a critical period for socialization and habituation it has to be continually reinforced throughout the animal's juvenile period. In the dog this is from twelve weeks to maturity.

Let us consider a practical example of how this research affects the dog owner. A puppy, well-socialized with children until it is twelve weeks old, will require the socialization to continue until it is at least six months old, and preferably until it is mature, for the full benefits to be achieved. The same rule applies to a puppy who has been habituated to hearing traffic in the first few weeks of life but is then kept in a quiet rural environment until it is six or more months old, i.e. without periodic exposure and reinforcement it is likely to become fearful in the presence of traffic.

The most worrying result of a failure to socialize and habituate is that a dog fearful of a person or an object will naturally want to move away from it to maintain a safe distance. If flight is denied it the dog has three options: to accept the person or item; to continue to struggle to get away; and to try and make the person or item go away. In the last case the dog will often use aggression. As the dog is in a state of fear at the time we have come to call this behaviour 'fear aggression' (this is a common not technical term). A profile of the fearfully aggressive dog is one that shows aggression when it is on the lead, i.e. when it cannot escape from the stimulus it identifies as threatening, but the same dog will not show aggression when it is off the lead, i.e. when it can escape. Additionally, dogs who have a tendency to fearful behaviour due to a lack of socialization often become very territorial at home and in the owner's car.

Once the dog has started to use aggression in an attempt to make other parties go away, a mechanism of reinforcement comes into play that causes the level of aggression to increase in intensity. The dog finds that in most cases, whether it is on the lead, at home or in the car, its aggression seems to work so it uses the display with increased intensity and overtness. Why, from the dog's point of view, the aggression seems to work is easy to understand: people who come within its vicinity, causing it to bark, move away again. Actually they had no intention of doing otherwise, but to the dog they appear to have retreated because of its aggressive behaviour, so its confidence in the use of aggression increases. The treatment most dogs afford postmen and other people who deliver things serves as an example of this mechanism. Having approached the property these individuals go away again, and from the dog's perspective it seems as if they have been successfully chased off. The dog is unable to realize that they did not want to enter the house. The more this event is repeated the more overt the dog's aggression becomes. Significantly, most of the dogs referred to me who are fearful because of a lack of socialization and habituation display aggression towards those things of which they are frightened.

Everything I have stated so far leads to the question of why, if the benefits of socialization and habituation are so irrefutably proven, are so many dogs undersocialized and habituated? The reasons vary, but an examination of the early history of the seventy-nine dogs I mentioned at the beginning of this chapter shows that they fall into two main categories (groups A and B):

A. Those that are retained by the breeder until they are well into, or even past, the critical period in an environment devoid of stimulation or with limited stimulation.

B. Those that are retained in the new owner's household until the puppy's vaccination programme is complete, often long after the critical period has passed.

Of those in group A we have to take into account the fact that breeders sometimes cannot find enough suitable homes quickly enough. Having said that, it is unfortunate that some breeders believe that most families are unsuitable to look after a puppy when it is six weeks old, although it is difficult to see what suddenly makes a family suitable when the puppy is eight, ten or twelve weeks old. All too often

Group	No. of dogs	Age acquired by owners	Puppy's environment
A	4	Up to 10 weeks	Barn or shed
	6	10–12 weeks	Kennel or outhouse equivalent
	16	12–16 weeks	Kennel or breeder's home
	15	Over 16 weeks	Kennel or breeder's home
B	38	6–12 weeks	Retained within new owner's home until vaccinations complete, often after 16 weeks of age

breeders, unaware of the harm they are doing, retain puppies well into and sometimes past the critical socialization and habituation period so that they, the breeders, have time to choose which puppy or puppies they wish to keep for showing before launching the rest on the unsuspecting public. There is in essence nothing wrong in a breeder retaining a puppy for as long as they want, as long as they systematically ensure that each puppy is properly socialized and habituated as an individual. Each puppy needs to learn to cope with the environment without the support of its litter brothers and sisters. Although this is possible, in practice, it is very time-consuming.

In group B, the implementation of vaccination programmes was a major contributor to the number of psychologically disturbed puppies. This was done in the name of the puppy's physiological well-being, which is essential, but in fact there is no need for owners to see themselves as having to choose between the needs of physical well-being and those of socialization and habituation. To understand why we need to know something of the principles behind and the history of puppy vaccines.

When a bitch has a litter she passes immunity (antibodies) to infectious diseases such as distemper and parvovirus via her placenta (10 per cent) and more importantly via her milk (90 per cent) to her puppies. The immunity puppies obtain from their dam gradually declines, and to stop them becoming vulnerable to infectious diseases the immunity is re-established by vaccination. However, when a puppy has an effective amount of the immunity it acquired from the dam, that immunity will cancel out any vaccination given, making it difficult for veterinary surgeons to know when to vaccinate.

In the 1950s a scientist named Baker showed that by twelve weeks of age 98 per cent of puppies have lost their maternally derived immunity to infection, which meant that if puppies were vaccinated at twelve weeks the vaccination would have a high take-up rate. To ensure that the puppies were not exposed to sources of infection in the meantime they had to be isolated in the owner's household until they were at least twelve weeks old, and normally for two or more weeks after that. This practice still lingers on, but although undoubtedly cost-effective, it completely ignores the puppy's psychological needs. Once again, it was Derek Freeman who pioneered the way forward. He had an urgent need to start socializing and habituating puppies within the critical period, i.e. from six weeks onwards, but of course he had to ensure their protection from infection. After consultation with one of the drug companies that produced puppy vaccines, blood samples were taken from bitches that were in whelp so that their antibody level (titre) could be counted. The lower the titre count in a bitch's blood sample the smaller the amount of immunity the puppies would gain from her milk and placenta. Consequently the actual time the puppies' vaccination programme should start could be predicted, and generally speaking it proved to be at around six weeks. Research elsewhere has confirmed that most puppies have lost an effective level of maternally derived antibodies by six weeks of age.

Guide Dogs for the Blind developed a policy of systematically vaccinating all puppies at six weeks and then repeating the inoculations at intervals to catch those few whose level of maternally derived immunity was too high for a vaccine to take on the first occasion. This removed the need to blood test every bitch for a titre count. The policy ensured that at any one time puppies were covered, either by maternal antibodies or the vaccine. In more recent years one drug company has recognized the need for early socialization and therefore early vaccination. As a result of their research they produced a vaccine which is specifically designed for early use, with the additional benefit of an ability to overcome what is known as the immunity gap. This is the period of time in which the puppies' maternally derived antibodies are too low in number to prevent infection but numerous enough to kill off any vaccine given, i.e. this type of vaccine will take as soon as the maternal antibodies are too low to resist infections. This lead has been followed by others and this sort of specific early vaccine is now readily available.

Whatever system of early vaccination is used, multiple or specific early cover, the principle has been proved safe and effective over the twenty-five years in which Guide Dogs for the Blind has successfully reared 24,000 puppies. Derek Freeman stated that he could count on the fingers of one hand the number of puppies that became unwell. Parvovirus, of course, being an entirely new disease that appeared in 1979, brought its own challenge as there was no protection, but now that vaccines have been produced to combat the virus, vaccination programmes starting at six weeks continue to prove effective in preventing mortality.

It must be remembered that all vaccination programmes have to be carried out under strict veterinary supervision and have to be boosted annually, even with the relatively fail-safe vaccination programmes I have been advocating. A puppy's safety can be further assured, if further assurance is required, by carrying it in a blanket until its vaccination programme is complete. The blanket will not only keep the puppy warm, but it will also prevent any toileting accidents causing the owner to have to put the puppy on the ground where it could be exposed to sources of infection, i.e. other dogs and locations where they have defecated or urinated.

When so much is known about the critical nature of early socialization and habituation, and there are proven effective early vaccination programmes, why are so many dogs kept in private homes up to and often beyond twelve weeks of age? The answers are tradition, money and ignorance.

The tradition of starting vaccination programmes at twelve weeks has been around for as long as vaccinations. Sadly, this tradition is changing only very slowly, and one of the reasons for the slow change is the fact that early vaccinations are more expensive. If someone is offered a vaccine at six weeks that will have to be repeated or a vaccine at twelve weeks requiring fewer or no repetitions, it is obvious which they would choose if they were unaware of the benefits of socializing and habituating puppies before they are twelve weeks old. When compared to the overall cost of feeding etc. the sum for extra or special vaccines is nothing, and will bring benefits throughout the dog's life. And it is not always the owners who are ignorant. Some breeders have amazing misconceptions about vaccination programmes and what should be done with puppies with regard to developing temperament, training and diet, and pass them on to new owners who accept the advice as if it were written on tablets of stone. The consequence is that many pet dogs fail to benefit from twenty-five years of progress and learning. Derek Freeman firmly believed that the programmes he devised were advantageous for both guide dogs and pet dogs alike. Although he is no longer with us his crusade goes on and one day all puppies will get the best start in life.

Having looked at the theoretical aspects of early socialization and habituation, what are the actual mechanics required to achieve it?

Instead of socialization and habituation being a haphazard affair with experiences occuring at random, as is so often the case, the puppy's exposure to environmental stimuli should be as systematic as possible to ensure the best chance of it developing a sound temperament and capacity to cope in all circumstances. A lot of responsibility lies with the breeder. Of course, it is the breeder who selects the genetic make-up of a dam and sire best suited to produce puppies of good temperament. Having said this, it is not known what percentage of a dog's, or even a human's, temperament is determined genetically and what percentage is determined by environmental influences. The breeder's role continues the moment a puppy is born, as it starts to get used to being handled and to the breeder's scent. The routines that are normally used to assist in whelping are enough to accomplish this, much more may distress the bitch. As the puppy and its litter mates grow up, the breeder should increase the amount of interaction the puppies have with them and other people. If the breeder is a woman, for example, and she is the exclusive, or almost exclusive human contact the puppies have, they are likely to be less well adjusted towards men and children. It is sensible therefore, to invite men and children into the household to see and handle the puppies, particularly if the puppies remain with the breeder after they are six weeks old. It is, of course, important that the veterinary surgeon's advice on hygiene procedures is sought.

Taking the trouble to ensure early and comprehensive socialization is in the breeder's own interests. So many lady breeders, for example, complain that their dog 'will not show under a male judge' because it is, to some extent, fearful of men through a lack of socialization with them, which results in apprehensive behaviour at the best of times, but when a male judge is in the show ring staring at the dog and attempting to touch it there need be little wonder that it cannot cope. I have, over the years, seen some extreme examples of this problem. In one case, a Great Dane puppy called 'Hamlet' was reared by a breeder who lived on her own and whose friends and visitors were all female. Hamlet was subsequently sold to an elderly lady who lived on her own in a rural environment. By about twenty weeks of age Hamlet's owner, not surprisingly, could no longer cope with such a large puppy, whose antics caused such a drama, so Hamlet was rehomed with a young family in a city suburb. I saw him two weeks later. He was able to cope with the mother of the family's presence perfectly and he was not too nervous with the children, who were about five and seven years old, but he was absolutely petrified of the father. As soon as the father walked into a room in which Hamlet was he would hide under or behind the nearest piece of furniture. If at any time the father came too close, even if he was just walking past, Hamlet's head and neck would appear and snap at him like a conger eel before retreating. It was also impossible, without treatment, to get Hamlet to walk out of the garden and around a quiet housing estate as he simply collapsed in a nervous heap.

It is important for breeders not only to socialize comprehensively the puppies in their care, but they must ensure their exposure to environmental stimuli. Not being able to take puppies off the premises in the first six weeks is limiting, but a puppy that has had regular experience of a television, vacuum cleaner, etc. will be more able to cope with the world than the one that has been shut away in a quiet kennel or room. Audio tapes of environmental stimuli can be made and played to the puppies. This may help but because of the dog's acute sense of hearing, tapes will never be a substitute for the real sound. However such techniques can be helpful if an older puppy is unwell or for some other reason cannot be taken outside the home.

Obviously, good breeders will make themselves responsible for acquainting new owners with the principles of socialization and habituation at the same time as they advise on approved diets etc.

It is a good idea for breeders to ensure that the prospective owners have enough time and dedication to continue the socialization and habituation process properly, because if they don't and the puppy subsequently develops a less than sound temperament it is the breeding and not the rearing that is likely, but not justifiably, to get the blame.

Prospective owners can maximize their opportunities to socialize and habituate their puppies by obtaining them at six weeks old, having already made arrangements for the appropriate vaccination programme with a veterinary surgeon. Of course failing to obtain a puppy at exactly six weeks does not automatically lead to disaster, but the later puppies are acquired the more precious time will have been lost and the less likelihood there is of developing a sound temperament. This is balanced to some extent by the level of sensory stimulation the breeder has provided. Basically, a puppy obtained from a chaotic, noisy family household is far less likely to develop a fearful temperament than one that has been kept exclusively in a kennel or farm building. Conversely, if a puppy has been raised in an environment devoid of stimulation, such as in a barn or quiet kennel, the comencement of socialization and habituation at six weeks does become critical.

The prospective purchaser of a puppy can check that some degree of socialization and habituation has taken place. Ideally, they will have sought out a breeder who will let them see the puppies with their dam in their living quarters prior to the optimum 'go home' age of six weeks. Searching questions should reveal the breeder's awareness of the need for a puppy's environmental enrichment, but the proof of the pudding is the reaction of the puppies themselves. They should appear to be content and confident. A few simple tests, such as the clapping of hands, the dropping of car keys etc. will enable the prospective buyer to gauge how well habituated the puppies are by observing whether they move away from the sound or towards it to investigate it. A mild reaction to the sound and a quick recovery from the surprise is ideal. Most telling of all is the puppies' response to the presence of strangers, i.e. the prospective purchasers. They should be willing to approach and investigate the newcomers and be happy to allow themselves to be handled.

Prospective owners should also observe the behaviour of the dam and any other dogs that are in the vicinity of the puppies. If the puppies have grown up in the company of a nervous or aggressive dog they may have learnt to be fearful or aggressive from its example and it may be wiser to look elsewhere for a puppy than to take the risk. Many guides have been written on how to choose an individual puppy from the litter; the subject is somewhat beyond the scope of this chapter. However, I would recommend that several points of view are sought and that potential owners are ruled by their head and not their heart. The heart can take over once the choice has been made.

If you are about to purchase a puppy, or as a breeder you intend to run-on a puppy, there is a lot you will have to consider, such as toilet training, preventing chewing and how to get a good night's sleep. In the turmoil and upheaval time must be put aside to consider the best locations for socializing and habituating your puppy. These guidelines should help:

Things that can be done at home

Visitors. Accustom your puppy to lots of visitors of both sexes and all ages. This will develop its social experience and help to keep territorial behaviour to manageable levels in later life. Ensure your visitors only say 'hello' and 'fuss' your puppy once it has got over its initial excitement so as to prevent the development of boisterous greeting behaviour.

Children. Accustom your puppy to being handled by your and/or visitor's children, but don't let them pester it or treat it as a toy. Remain in a position of supervision. Arrange to meet someone with a baby regularly, especially if you plan to have a family. This will help to overcome the common worries about how the family dog will react to a new baby and toddlers.

Feeding. Accustom your puppy to you and other members of your family adding food to its bowl when it is eating. This will teach it that you are not a threat and prevent the development of aggression over food when it is older. Conversely, teaching your puppy that you can take its food away when it is eating is a bad idea as this approach can cause the development of defensive behaviour later in life.

Grooming. Groom your puppy every day, even smooth- or wire-haired breeds who may not seem to need it. Grooming will accustom your puppy to being thoroughly handled and coincidentally it will help prevent the development of dominant behaviours.

Veterinary examination. Every day examine your puppy's ears, eyes, teeth, lift up its feet and check its paws and check under its tail. When your puppy is happy about this get other people to do it (it makes a good talking point at dinner parties!). The purpose of the exercise is to accustom your puppy to veterinary examination, very important, especially if first-aid ever has to be administered.

Domestic sights and sounds. Expose your puppy to domestic stimuli such as vacuum cleaner, spin drier, etc., but don't make an issue of them. The puppy should get used to them gradually without being stressed.

The postman, milkman, etc. Carry your puppy and meet these people as often as you can. If your puppy gets to know and like them and more important learns that they will not 'run away' if it barks, it is far less likely to show territorial aggression towards them when it grows up. (Many householders have to collect their post from the sorting office because the postman will not deliver as a result of their dogs' behaviour.)

Cats. If you have one introduce your puppy to it. Keep the puppy under control and reward it for not pestering. Be careful not to worry the cat, as it may scratch your puppy. Placing the cat in a cat carrying basket just out of the puppy's reach can be a useful method of introduction with little chance of an unpleasant incident occurring. This can be repeated after a few days so that both puppy and cat learn to become settled in each other's company.

Other dogs at home. If you already have a dog introduce your puppy to it in the garden. Once the initial acceptance has been made by the older dog, the two should find their own level and settle down without too much intervention from you.

Prevent play-biting. In pack society once puppies become active they play physical games with each other and pester the adults by pulling their ears, tails, etc. In the early days puppies have licence to do what they like but as they grow up, adults and litter mates alike become increasingly intolerant, especially of their very sharp teeth. By eighteen weeks puppies learn that hard-mouthing or play-biting is taboo and a reprimand will quickly follow any transgression of the rules. When a puppy is introduced into the family this learning process is normally incomplete. The family must take over where the puppy's mother left off.

How is this done?

Whenever a puppy uses its teeth in play the person concerned should respond with a sharp 'No!' and sound as if they have been really hurt. Then walk off and ignore the puppy for about five minutes. In this way the puppy learns (a) to limit the strength of its bite in both play and for real and (b) that biting is counter-productive as an attention seeking device.

Leash training. Prepare your puppy for walking on the lead by getting it used to its collar and lead in the garden.

Going solo. Socialization is very important, but so is learning to be alone. Puppies who are not accustomed to being left unattended on a regular basis are much more likely to suffer from separation anxiety (i.e. become anxious when separated from the owner) in adulthood. The three main symptoms of separation anxiety are destructiveness, incessant howling or barking and loss of toilet control.

To help prevent your puppy from suffering from this very common syndrome, you need to leave it unattended (i.e. in the house on its own) for over an hour on most days, preferably in the area that it sleeps in overnight, which should not be your bedroom, as sleeping there can contribute to separation anxiety and other problems.

For your puppy's safety, to prevent it from toileting in inappropriate places, chewing inappropriate items, etc. ensure its area is 'chew proof' and free from hazards such as electrical cables, etc. You may need to construct or buy some purpose-built barriers to make a pen. Indoor kennels are often used and are readily available. Leave your puppy with some appropriate chew items, such as long-lasting chews from the pet shop, and fresh water.

Initially you should accustom your puppy to you sitting in another room, with the door between you open. Over a period of time the routine can be carried out with the door shut. Once your puppy accepts this you can start to leave the house; go next door for a coffee, for example. Gradually extend the time you are away until you are absent for over an hour on a regular basis. Do not go back if you hear your puppy crying. Return when it is quiet. If a puppy thinks it can 'call you back' it may never accept being left.

Be very matter of fact about going out and coming home. If you fuss your puppy before leaving you will unsettle it and make it want to be with you at the very moment you want your absence to be accepted. (There is nothing in dog language for 'Bye-bye, see you later'. Any interaction means, 'Let's go!') Too much fuss on returning home highlights the loneliness of your absence.

Things to do away from home

Go to all the environments you can think of that will help your puppy become 'bomb proof'. Start in quieter places and gradually find busier ones.

The street. Expose your puppy to the sound of traffic and the movement of people. Start in quiet side streets and gradually build up to busy ones.

Places where people congregate. Any environment where people tend to congregate to sit and chat will do, so that they have the time to take interest in and handle your puppy.

Children's play areas. These are obviously a good place to meet lots of children (but consult your veterinary surgeon about the appropriate worming programme before bringing your puppy in contact with children). Children should not talk to strangers, so make arrangements with their mothers. Start with just a few children and control their enthusiasm to prevent your puppy from being overwhelmed, which can easily happen.

The car. Plenty of car travel will accustom your puppy to it and help prevent car sickness. Do not let your puppy sit on the front seat or on someone's lap. Accustom it to travelling in the place it will occupy when it is an adult.

The countryside. Accustom the dog to the sights, sounds and smells of the countryside and livestock etc. (in your enthusiasm don't forget the Country Code).

Leash training. Once your veterinary surgeon has said that your puppy can be safely walked on a lead instead of carried, carry on as before but go back to using quiet areas, then gradually build up to noisy and busy ones again. In addition think about the unusual places to which you can accustom your puppy, for example, open staircases can be a problem, as can the vibration of station platforms when trains arrive or the movement of the floors on trains, buses and lifts. In the countryside keep your puppy on a lead and reward it for staying with you and ignoring livestock.

Socializing with other dogs

Removing a puppy from its dam and litter mates at six weeks is ideal in terms of socializing it with people but its socialization with other dogs stops. As already discussed, socialization will wear off, which means that some steps have to be taken to ensure that the process of learning to interact with other dogs continues if owning a maladjusted puppy is to be avoided. However, socializing with other dogs does not entail allowing your puppy to run amok with dogs in the park. If they, the other dogs, are not properly socialized their own interactive and communication skills may be poor, which can often result in a misunderstanding and aggression. This sort of encounter could result in the puppy learning to be aggressive towards other dogs. If you go to any town park on a Sunday afternoon you will see plenty of dogs not getting on simply because they cannot communicate properly.

In order that their puppy's canine interaction skills can be properly developed, it is very important for puppy owners to locate and attend one of the increasingly popular puppy socialization classes, even if it means travelling some distance to get there. Sarah Heath's chapter discusses these classes in more detail.

Finally, what should you do if a puppy shows fear whilst it is being socialized/habituated?

(a) Do not overreact. If you try to reassure a puppy it may reinforce its fear, as it will see your reassurance as your fearful response to the thing that frightened it. As 'pack leader' you should appear to be unaffected and unworried so as to 'set an example'.

(b) Don't try to pressure a puppy into approaching the item as you will highlight its fear by drawing its attention to it.

(c) Expose the puppy to the type of stimulus that worried it as often as possible, but initially from a distance (i.e. reduce the size of the stimulus) so that the puppy can become desensitized to it. As the puppy's reaction improves you can gradually increase the amount of stimuli.

(d) Reward the puppy every time it does not react to the stimulus, or as soon as it recovers from its fright if it does react.

4 Understanding Aggression

Robin Walker

Aggression and fear in a predator
....

Whether you are sweeping off to a 'top table do' in white tie and tails for quail's eggs, claret and a discourse on everyone from Aristotle to Wittgenstein; off to the pub for a pie and a pint; or sitting by the front door with your lead in your mouth, your prime directives are *food* and *sex*. And aggression enables a creature to gain nourishment and reproduce itself.

No species of social animal demonstrates the complicated balance between fear and aggression, and its role in survival activities such as feeding and breeding, so vividly as the spotted hyena. Although the hyena is a ferocious predator on other species it has more to fear from its own kind than other hunters. To acquire food the 180-lb, bone crunching hyena must co-operate in large groups, which mark out their territories with faeces and anal gland pastings. The clan territories are ferociously defended, invading hyenas being killed or pursued for up to two miles. Hyenas have been observed to flee the faecal markers of a rival group and hunting groups may abandon a hunt on entering another clan's area. On finding itself in 'enemy' territory, a hyena's body posture will betoken fear (ears flat, mouth open, tail between legs, hindquarters down), which increases the further it penetrates the foreign range. This fear however gives way to aggression if the hyena is accompanied by such numbers of its own clan as to outnumber the opposing group. Ferocious fights ensue and the vanquished are eaten.

To ensure the continuance of the clan the females must travel prodigious distances with the males, co-operate in the kill, consume enough meat to maintain their own strength and produce nearly ten pounds of milk in their huge udders, and then run back to suckle the cubs. They must also defend their offspring against lions, nomadic scavenging hyenas and the cannibalistic members of their own clan. Not surprisingly the females are larger than the males, dominate the feeding at the kill (shouldering aside males and lower ranking females) and supervise the survival of the fittest offspring. To achieve this the females have levels of male hormones comparable to the males and bear imitation male genitalia (enlarged clitoris and false scrotum). Drenched in male hormones in the womb, the cubs emerge from it with open eyes and erupted milk teeth and attack each other ferociously. Some are killed by the instinctive lethal neck-shake, others die of starvation having been bullied from the teat. When not within the breeding den, itself too small to admit adult hungry hyenas, or in the group crêches or communal dens found in some clans, the cubs have been observed creeping out of reach into small dens they have dug for themselves.

The spotted hyena demonstrates a complete spectrum of fear and aggression strategies. On the one hand we see ferocious organized predatory aggression fuelled by high levels of male hormones, on the other precipitate flight from other hyenas with righteous territorial authority, from lions at a kill and from elands in a tight defensive ring with their horns offered to the attackers. Should lions attack the den area, the females will summon help by characteristic whooping and conjoin to drive off the feared predator. A social hierarchy is maintained by lethal aggression against interlopers and the cubs of rival mothers but a highly developed system of appeasement body language, whooping and genital exploratory bonding behaviour reduces the mortality rate within the clan.

The programmes controlling the mental and physical mechanisms of fear and aggression, 'fight or flight', are closely interlinked in the brain and the body chemistry is almost identical. We should not therefore be surprised that it is possible for an animal to switch in an instant from predominantly fearful behaviour to overtly aggressive behaviour, as when an animal frozen in fear (tonic immobility) suddenly attacks its aggressor. Similarly, a predator racing in for the kill may veer away (avoidance) if the prey suddenly turns, lowers its horns and charges. An organism's response will depend on its breeding, hormones, what it has learned and the powerful commands of its social programming. The description of a specimen situation is rarely typical. We are presented with something of a 'pick and mix' or mosaic of fear and aggression which is modified by the surroundings and the circumstances.

The ancient 'blueprint' or programme for the dog is the social group or pack organized for getting food and producing offspring. To get the food the pack must co-operate. This requires a leader and a group whose members all sleep at the same time and are available to arise and hunt at the same time. To maintain the vigour of the species the leader will mate with the 'best' females and the production of 'inferior' offspring will be discouraged. This programme remains more or less intact. Little has changed in respect of food; the instinct to hunt and kill is still strong; as is the need to guard the territory where the food is found. The dominant urge to organize these behaviours is also still strong. The reproductive instincts have blurred but surface in inter-male conflict, inter-bitch conflict of lethal intensity and the teeth chattering dissonance of my terriers when they scent newborn pups in the clinic. They are powerfully prompted by something from the past to kill the upstarts. Something powerful in the present tells them they had better not!

Aggression: the neurophysiology

The greater part of the very little that is known about the mammalian brain is incomprehensible to us. Experiments with electrodes, accident victims and surgically assaulted brains give vague information as to what parts of the brain might be capable of, but not of the complex pathways and circuits for receiving and analysing threats and startles and orchestrating the finely balanced responses and the split second changes of strategy. These damaged brains are like an immensely complex telephone exchange into which someone has tossed a fragmentation grenade. A passing ignoramus is then given a huge bag of small change and directed to a basic public telephone with instructions to find out what damage might have been done! The result is an incomplete and probably misleading idea of the nature of the harm that has been done. However, there is also patient and less violent research under way into the amazing variety of chemical messengers and powerful hormones produced by a body in a state of fear, anger or stress and this is beginning to yield helpful information. This information is arriving at the rate of 1,000 articles a year and one book every two months.

One paragraph in *Fears, Phobias and Rituals* by Isaac Marks, one of the best books on the subject, says it all. Here, paraphrased and translated into intelligible terms is the basis of most of our problems in life!

Fear and stress are accompanied by widespread endocrine changes. There is release of controlling hormones, cortisones, adrenalin and noradrenalin, growth hormone and alterations in testosterone levels. These hormonal responses are more marked during acute stress than long-term stress and are soon set at steady levels. The hormone responses are often at cross-purposes with other body systems dealing with emotional response. The cortisone response occurs especially with sharp changes in external (or internal) stimulation, and with novelty, uncertainty and conflict; it varies greatly according to how often the stress happens, how long it lasts, sex, species and social rank. *Ability to mobilize and demobilize adrenalin quickly may relate to good adjustment and parallels the more rapid return to resting levels of nervous responses to fear in normal than in anxious patients.*

Stress can increase susceptibility to infectious diseases and cancer, presumably through suppression of the immune system. Fear in dogs raises the red and white blood cell count. Blood fats rise with short-term emotional arousal, and myocardial infarction recurs less often if patients reduce time-pressured behaviour. Free fatty acids in the blood may rise more with anxiety than hostility.

'Mad dog' attacks family: a case history

Twenty years ago I was called by the police at 2 a.m. to a household whose dog had attacked the family. They were clients of mine and their previous dog had been destroyed for attacking them. The breed of their current dog was not known. I collected a very large, brave RSPCA inspector, who was a personal friend, and drove to the house. The family were milling around in the garden in a state of great excitement. They numbered around eight, mostly children, and were in various states of undress, here a shirt with no nether garments, here pants only. None of them were hurt. All of the ground floor windows and both front and back doors were open, the family having exited via every available opening.

Armed with torches (the incumbents were in difficulties with the Electricity Board), large grasping nooses and gauntlets we crept cautiously into the house. The inspector insisted I keep behind him and I was more than happy to comply. We finally discovered the hellhound sitting on the second tread of the stairway with drooping ears, a petrified expression and not a sign of aggression. It being about the size of a small Corgi, the inspector scooped it up under his arm. As no sense could be obtained from the owners, who represented as clear a case of familial insanity as I have ever encountered, the police inspector asked for the dog to be destroyed in accordance with their wishes. I told them that they had driven their previous dog 'mad' and should never have another, or if they did to keep well clear of me. I told the police inspector that the dog should keep the house and he should take the ******* family away!

It is my opinion that a perfectly normal dog had been subjected to the *novelty, uncertainty, conflict, short-term and long-term stress* that only a chaotic, manic maelstrom of shrieking barmies could provide and sustain. He had obviously attempted to dominate them in the interests of social order.

By and large the uninformed, instinctive dog owner (or parent) 'trains' his dog (or 'brings up' his offspring) by shouting and waving his arms angrily. It is a tribute to the natural goodness and anxiety to please of the average puppy (or child) that this extraordinary and universal standard of training has the reasonably good results that it does. Those with more exacting standards of behaviour and manners would possibly say that it is surprising that matters are not very much worse. The trouble really begins when the human who 'instinctively knows what to do' is confronted by the disturbed personality. The normal range of aggressive behaviour can be addressed by establishing that the dog owner is dominant and by diverting predatory aggression into controlled games. The greatest difficulties are associated with fear and aggression which go 'off the scale'. Phobia (extreme or irrational fear) and rage (extreme aggression) are truly psychopathic, having no survival value.

It needs little thought to predict the fate of an anxious or depressed hyena, or one who goes for lions or is oblivious to scent markers.

Rage
Gizmo: a case history

In March 1989, Gizmo, a Lhasa Apso-Poodle cross of five months was flattened playfully by the family German Shepherd. He was concussed and had a chip fracture of the shoulder. He recovered uneventfully. Two months later he had a series of violent rages. The first was in the cab of a horsebox on the way to a show. He had a second whilst tethered at the show and a third in transit to a vet that evening, where he was unexaminable. In the morning he was admitted to my clinic for observation. He seemed normal for a couple of days but was heard growling to himself on occasions. Gizmo was deemed to be a case of low threshold aggression and received some Tardak to reduce his male hormone activity for a spell. His owners were advised about dominance and diet.

Gizmo's owners coped with him until August 1991 when it was decided to neuter him. This has since made him very much better, in his owner's opinion.

After three years Gizmo's owners are able to say that his 'rages' always occur when he has been sleeping. This suggests that he has hidden in his brain one of the slow-burning, spluttering, electrical activities which can give rise to full-blown epileptic seizures in other individuals. Veterinary investigators recognize rare forms of explosive rage. Attempts are made to organize symptoms under a 'disease' label. For example, a diagnosis of 'aggression caused by seizure activity' may be made if the following conditions are met:

1. The aggression does not fit a description of a normal species-typical behaviour.
2. There are concurrent neurological or pathological signs, such as abnormal electroencephalography readings or actual fits are seen.

3. Drugs which normally sedate aggressive dogs cause fits or aggression.
4. Anti-epileptic drugs actually suppress the aggressive behaviour.

The medical literature on human aggression is fascinating. It is possible to include in the general definition of 'impulsive disorders of conduct' or 'personality disorders' everything from being 'ratty' first thing in the morning to psychopathic rages associated with grand malseizures.

Dogs who are tired or stressed, by the unpredictability of healthy and boisterous families, for instance, suffer from plain, ordinary irritability. Like humans, they can often be very irritable on waking and at the end of stressful working days. As to the cause, there is a well-established correlation between mood and blood sugar levels, although this is not easy to verify in dogs. Few things can be more difficult to accomplish than taking the brain wave recordings and hourly blood sugar samplings of seriously angry dogs. The scientific proof of much of what lifetime dog handlers know intuitively is simply too dangerous to collect. Dogs do however need blood sugar, and meat is a particularly lousy source if you need it quickly, and moreover if you are only fed once a day there is an awfully long time to wait for it. Startling improvements in dog temper can be achieved by switching them to slow-release carbohydrate diets which give a more steady blood sugar level.

Much of the perceived 'well-being' of vegetarian converts is due to better sugar levels and relief from the stupefying onslaughts of sleep-inducing chemicals triggered by large intakes of protein. So remember to breakfast well before meeting the family in the morning, eat a substantial carbohydrate meal at lunchtime, take a small meal in the evening – and never kick the dog's basket when occupied.

Violent outbursts of rage with or without gross disturbance of consciousness are the psychomotor form of a fit. These patients are liable to periods of hours or even days of increased irritability with outbursts of rage triggered off by *comparatively trivial stimuli*. These individuals will not necessarily ever have a fit. There may, however, be relatives who have suffered seizures.

American workers have described a group of dogs that were hyperactive, had a short attention span, were difficult to train, tended not to respond to tranquillizers, were destructive and aggressively resisted restraint. Careful screening of these categories will narrow them all to a very small number of genuine pathological cases. The combination of genetic drives, diet-influenced physiology and learned responses must be addressed by the therapist, and it is dangerous to overlook the possibility that there is disordered functioning. A dog which suddenly awakes in a state of clouded consciousness ('glazed eyes') and launches an attack in response to a very minor stimulus may well be having the canine equivalent of a temporal lobe disturbance. Mercifully cases of psychopathic rage are very rare. The majority of problems arise from a conflict between the natural drives of the dog and the lifestyle of the owner. It is possible that the combined observations of alert puppy socializers and behaviourists will discover learning difficulties in the young dog similar to those now recognized in children. Problems with perception, sight, hearing and understanding bring many children to frustration and conflict with exasperated adults who are ignorant or intolerant of such disability.

Current views on aggressive behaviour

Aggressive behaviour has multiple causes and any study that attempts to identify a single, all-important cause is bound to be a simplistic one. A recent study has found that the conditions related to the use by parents of harsh punishment were in the following order of importance:

(a) Perception of the child as difficult to handle.
(b) Proneness to anger on the part of the parent.
(c) Rigid parental power assertion.
(d) Intra-familial problems and conflicts.

There is an incontrovertible link between harsh parental punishment and anxiety and helplessness. There is a style of parenting, teaching and training which I describe to my clients as follows:

> Harsh training is rather like attempting to alter the shape of slippery bars of soaps by treading on them. Some bars may be shaped to your satisfaction, sadly a few will be smashed and a large number will escape from underfoot at great speed and in totally unpredictable directions.

Human aggression has been divided into categories together with responses to 'bad' or 'deviant' behaviour. Four categories have been useful in the studies of children:

Category	Definition	Likely exclamations
Specific or instrumental	Concerned with obtaining or retaining particular objects, positions or desirable activities.	Gimme that. It's mine. I wannit!
Teasing or bullying	Directed primarily towards annoying or injuring another individual, without regard to any object or situation.	Dunno. She asked for it. Boys will be boys.
Games aggression	Playful fighting escalates to the deliberate infliction of injury.	Ow! Take that!
Defensive aggression	Provoked by the actions of others.	He hit me first, miss!

Children will interact with the family pet in all of these ways. Adults will all too frequently approach the pet in the *childish* mode and respond as children.

Criminal violence by adults has been classified as follows and can be applied to aggressive 'training' in any context.

Category	Definition	Likely exclamations
Instrumental violence	Motivated by a conscious desire to eliminate the victim.	Either you kill 'im or I will!
Emotional	Impulsive, performed in extreme anger or fear.	Christ he's bitten me! Oh you little bastard! Take that!
Felonious	Committed in the course of another crime.	I've a tenner on the terrier and it's safe 'cos I broke the badger's jaw first.
Dyssocial	Violent acts that gain approbation from the reference group and are regarded by them as correct responses to the situation.	I am sure, madam chair, that you will agree hanging is too good for them. They need a sound thrashing. Hear! Hear!

Karl: a case history

Karl was just over a year old when he suddenly became aggressively possessive of his food in the tiny kitchen of a time-share apartment. His owners' response was righteous wrath, which produced such remarks as the following, 'Don't you growl at me. Right, give it here. Into the bin with it. That will teach you.' In the ensuing months the dog continued to growl over his food, but it was now also occasionally elicited by approaching his toys, touching him or moving suddenly on the furniture. Upon growling he would always retreat behind the furniture. After a time he would emerge with his 'bone' and ask to play. Occasionally he would 'freeze', push his head against his owners and then growl. He had not bitten any of the family.

Karl was not in the least defensive about the home, greeting visitors boisterously, rearing up to lick them, thrusting his nose into their groins and pawing at their legs. During the consultation Karl's body postures were submissive except when he was invited to play. He liked tugging and pulling games and was very determined to win. When his male owner entered the room he slunk behind the wife's chair. The man tended to use a very rough, aggressive tone with the dog, and while he was out of the room talking to a neighbour, his wife and daughter affirmed with considerable vehemence that he was aggressive towards it.

Out of the house Karl's behaviour was normal except that he was apprehensive of other dogs and dark shadows or black plastic bags. At the local GSD training class he was very much the 'top dog' but his threatened aggression toward the other dogs turned to play if he was allowed to approach.

The owners had already obtained copies of *Think Dog* and *The Good Behaviour Guide* and were motivated to attempt behaviour modification. It was decided to establish a clear hierarchy in the family with the dog ranking below the daughter, who had no influence on Karl at all. The full dominance-reducing regimen of controlled games, ignoring of attention-seeking behaviour, etc. was instituted. The food guarding behaviour was to be extinguished by the invincible *two bowl ploy*. This technique involves placing an empty bowl before the dog at feeding time. After he has searched it and is looking up in puzzlement a handful of food is slowly placed in the bowl while the dog is admonished to wait. The hand (and face) are withdrawn and the command 'take it' given. Slowly, handful by handful, the dog comes to learn that the owner is the source of food not a competitor for the meal.

No progress was made over the next month. Indeed, the training class had to be abandoned because Karl began to challenge his owner during the work whilst regarding the trainer with rapt attention. Something was going wrong.

Tactful questioning of the wife revealed that the owner was a strict authoritarian within the family, especially towards the daughters. Even more tactful discussion with the man himself revealed that he had not wanted the dog in the first place. He was, in addition, very unhappy at his work and would usually come home in a sour frame of mind.

Karl would greet him with a tremendous display of anxious submission seeking to be reassured of the dominant individual's protection. Instead he was rejected by the owner who saw this as just another nuisance. As a result the dog was beginning to show increasing fear with sub-dominance. The man went further in his treatment of Karl. On entering any room he would look round, locate the dog and stare at it to see if it was going to offer aggression! The rigid, intense application of dominance techniques by this authoritarian (who did not want the dog) was like something from a Victorian novel of childhood! The nervous food guarder had to endure intensely protracted mealtimes in the presence of an unrelenting disciplinarian.

In my experience few things are more dominant than ignoring the dog and regularly forgetting to feed him! Such owners are adored and those who notice and care and nurture are unable to persuade the dog to do anything!

My advice was to greet the dog effusively and thus reassure him of his safety in the pack and take him for walks and play his favourite games from time to time and otherwise ignore him. As a result his behaviour on walks improved enormously. He is calmer now and there is much less growling in the household. It is impossible to fake affection for dogs, children or troops. If those under your command sense that in general you hugely approve of them they will accept stricture. If they sense that you despise them there will be mutiny. Increasing repression just makes matters worse.

Aggressive therapy

The therapist must constantly remember that, however well analysed the behaviour may be in relation to the existing environment, the intrusion of the therapist will change the situation. Aggression from the veterinary surgeon or behaviour therapist will cause either fear or defensive aggression or both. For instance, if a puppy is seized roughly and held immobile whilst afraid it will see itself in the situation of having been seized by a predator. Extreme fear may pass through tonic immobility (apparent submission) only to become defensive aggression as it is released (apparent escape). A false lesson may be learned, i.e. aggression terminates the consultation.

The fearful dog that is likely to resort to aggressive defence should be carefully medicated with tranquillizers prior to handling. Frequent handling programmes should be set up with reducing medication.

False lessons

In some senses the sociology of the hyena clan or wolf pack is an unfortunate model for social reform, because a superficial study of the 'alpha' wolf and its control of the pack is likely to comfort the aggressive trainer. Sadly, the dominance hierarchy does at first sight seem to be a bully's charter: 'It's got to be the way. It's what they understand. I mean, your hardest fighter is top dog. Just show them who's the boss, that's the way.'

Sometimes such a philosophy meets its nemesis when the 'alpha' trainer triggers a fatal attack when he trips in the feeding pen, whips out a handkerchief unexpectedly on a walk or suddenly sneezes in front of the TV, but more often it is family members or passers-by who suffer the attentions of the aggressive dog in the absence of its dominator. It is not amusing, to say the least, for a wife to spend an entire day, from the moment the macho 'alpha' husband leaves for work to the time he returns from the pub, trapped in the bathroom (or in my worst case pinned against the hall wall by a large dog with no sense of time).

There is no place for violence in the management of dogs, children or horses, yet these three species are the subject of sustained punitive aggression by invincibly ignorant individuals from the lowest to the highest strata of society.

The behaviourists have much to offer animal and human society and the gathering and presentation of incontrovertible evidence of the appalling harm done by impulsive punitive training is a major part.

5 Preventing Aggression

John Rogerson

Recent trends in dog ownership tend to suggest that the average size of the family dog is on the increase. This change highlights several problems of control that traditional training techniques have failed to resolve. The owners of smaller breeds, whilst they may still have problems, at least do not have such *big* problems! Many traditional dog training methods rely on teaching control using such equipment as leads and choke chains. But if you take a ten-month-old dog that weighs over one hundred pounds, put a choke chain around its neck with a leather lead attached and then try to drag it around using your own strength there are only four possible outcomes:

1. If you are an extremely fit and active person you may indeed manage to impose your will on a reluctant to oblige dog.
2. You will tire much more quickly than your dog and you will have taught him to use his strength against you and to be persistent in order to get his own way.
3. Your dog will begin to lose any respect that he may have had for you as a result of your obvious failure to impose your will against him in a trial of strength.
4. Realizing your comparative weakness your dog will take on the responsibility of pack leader and see you as a member of *his* pack instead of the other way around. Whenever there is any confrontation between the two of you, your dog will begin to use aggression to resolve the misunderstanding.

We therefore need alternative methods of putting the owner firmly but gently in charge of their pet that do not require the physique of a bodybuilder to be effective.

What is control training?

The dictionary definition of control is to command, direct, rule, check, limit or restrain, but some trainers/owners seem to take it to mean to shout, smack, dominate, push, intimidate or threaten. Such

52

'training' is usually started at around the age of six months, which is often when problems have already begun to develop. Owners are often advised to wait until this age because the dog would be unable to take any hard, physical correction before then.

To me control is simply the owners' ability to assert their influence over their dogs in such a manner that there is mutual co-operation between the two parties. With this type of 'behavioural training' it is possible, and desirable, to commence training just as soon as you get your puppy. If you obtain your pup from a reputable breeder then it is possible that this sort of training will have already started.

In order to understand how behavioural training works let's look at the experiences of two blue Great Dane puppies out of the same litter and see what happened to each of them during the first six months of their lives.

Rio and Roger were brothers and both left their litter on the same day at the tender age of eight weeks. Both went into new homes where there were husband, wife and teenage son and where there was already an older dog of a different breed.

Roger's tale

Arriving home, the new puppy was put on the floor of the living room to meet Sam, the five-year-old crossbreed. Roger was a little apprehensive at first but after a while he decided that this other 'strange' dog was all right to play with and so there were short bouts of play interspersed with lots of exploratory behaviour in this new environment.

After two hours had elapsed Roger found himself bursting to go to the toilet and so he left the room and went into the hall to find a suitable spot. He quickly located a suitable area at the bottom of the stairs, relieved himself there and went back into the room to play with his new friend Sam. During the next thirty minutes, although Roger did in fact go over and briefly interact with the human element of his new pack, he was already beginning to learn that it was more fun to play with other dogs. This was brought home to him in a dramatic fashion when, thirty minutes after the act, he was dragged unceremoniously into the hall, his nose was pushed into his own excrement and his owner's hand rained stinging blows on his rump. Roger was then growled at by his owner and found himself all alone in the back garden for what seemed like an eternity. He cried and scratched at the door and was finally allowed back in to play with his friend Sam.

Day two in his new home saw a much more settled puppy who had started to learn a great deal about life in his new surroundings. When he was brought into the house after a particularly energetic play session with Sam outside in the garden, Roger was lifted up to have his feet cleaned and dried. As he was lifted up he wriggled around and tried his utmost to escape and when he was restrained further he became almost hysterical and so was immediately put back down on the floor and given his freedom. His owner released him in the belief that he would become easier to handle and accept being restrained after being given more time to settle in.

Later that week Roger started to become more interested in playing games with his owners but such play sessions only ever happened when Sam was not in the mood to play with him. By the end of the second week Roger seemed a perfectly normal outgoing puppy who had settled into his new environment with few problems apart from a growing reluctance to be cuddled and restrained in any way. His behaviour with Sam was giving no cause for concern as the two of them seemed to be the best of friends. Roger would play with his owners but tended to become over-excited and nip and bite at their fingers, arms and legs. Telling him off when he was playing these games only seemed to make him more excited and tended to make him increase the severity of his bites.

As the days passed Roger became progressively more and more difficult to handle and the amount of smacking and shouting that seemed to be required in order to maintain some sort of control began to escalate. The first time Roger was taken out on a collar and lead was at the age of sixteen weeks, after his course of injections finished. This proved to be quite an event as he pulled in each and every direction except that in which his owner wanted him to walk. Exerting more pressure on the lead to make him walk correctly either made Roger rear up like a wild horse, occasionally biting at the lead in frustration or else he would sit and refuse to move at all.

Things really came to a head when Roger, at the age of seven months, was taking his owner out for a walk. He spotted a dog in the distance and desperately wanted to say hello to it, so he suddenly lunged forward taking his owner by complete surprise. Ignoring the threats and shouts directed towards him he proceeded to drag his owner bodily across the road, narrowly avoiding the traffic. When he reached the other dog and owner they both objected to his approach and whilst the other owner shouted abuse at Roger's owner, the other dog launched an attack which consisted mainly of noise and threats.

The next time Roger was taken out and saw another dog his owner was ready for him and at the first hint of disobedience he was physically and verbally chastised. He then became even more excited and frustrated at being restrained and bit his owner's arm.

By the time Roger reached ten months his owners listed the following problems with him:

1. He was becoming increasingly aggressive towards other dogs and

would take no notice at all of his owners' attempts to control him whenever he happened to spot another dog.

2. He was very difficult to control when out for a walk on the lead and the owners were becoming increasingly concerned for their own safety because of Roger's superior strength.

3. If anyone tried to restrain him physically he became hysterical and often aggressive. This was particularly embarrassing when Roger was taken to the vet to have an ear infection treated. He not only refused to allow the vet to restrain him for the examination but would not allow his owners to put the drops in his ears that had been prescribed to treat the condition.

4. If his owners attempted to chastise him for a minor misdemeanour in the house Roger immediately became aggressive.

5. Roger's behaviour with guests was becoming something of an embarrassment as he would not leave them alone. He either had to be locked away in another room or put out into the garden with Sam.

His owners realized they needed to obtain some sort of control over Roger, so, after seeing an advertisement at their vet's, decided to take him along to their local dog training club. . . .

Rio's tale

Arriving home, Rio's owners immediately took him out into the garden and allowed him to run around and sniff the ground and generally explore this new environment. It was not long before Rio went to the toilet and was then introduced to Ben the four-year-old resident Labrador. This introduction also took place out in the garden away from the house where Ben slept and had all his favourite toys.

Rio was a little apprehensive at first but after a few minutes of allowing Ben to sniff him, Rio decided that a short bout of play was the order of the day.

Both dogs were then taken into the house and Rio was allowed to explore this new environment to his heart's content. When Rio awoke after falling asleep for a short while he immediately ran towards the garden where he had relieved himself previously but found his access blocked by a door. His owners, realizing that he would want to go to the toilet when he awoke, quickly opened the door and accompanied him into the garden where he soon found his previously used spot. When he was taken back into the house he spent some time interacting with his new owners whilst Ben amused himself with a rawhide chew he had just been given in the adjoining room. When Rio was finally allowed to return to the room where Ben was there was a brief play session before he tired and lay down on the rug in front of the fire.

Day two in this new home saw much the same pattern as Rio settled into this environment with his new pack. It was after lunch that Rio was lifted up to have his feet cleaned and dried that the first problems were encountered. He was determined that he did not like this experience and began to struggle to escape. When he was restrained further he became almost hysterical but his owner, realizing that Rio should not get all his own way and be released whilst he was behaving badly, simply held him gently but firmly until all the struggling had stopped. He then spoke to Rio in a reassuring manner and continued to dry his feet. Rio quickly learned to be touched and handled on his owner's terms and, because this grooming ritual was carried out on a daily basis, soon became accustomed to this type of restraint and handling.

Later that week Rio became more interested in playing games with Ben but would usually only do so if his owners were not available to play with him themselves. Even during the most energetic games with Ben, if he was called away with the offer of a toy to play with, he would readily leave his canine companion to play with his owners.

Most games with toys seemed to involve chasing, pulling and tugging which Rio carried out with so much enthusiasm that he occasionally got carried away and grabbed playfully at hands and clothing. Whenever this happened a 'time out' was called to allow him to settle down before the game could continue. His owners also made sure that Rio understood the rules of these games and taught him to release the toy immediately when told to do so. At every play session the owner's control over Rio improved rapidly and the instances of over-excitement to the point of inhibited play-biting had decreased dramatically.

Rio slowly became accustomed to wearing a collar by having it placed on for a few minutes several times each day whilst he was playing or eating, thereby making the whole process a pleasant one. For the first few weeks in this new home Rio had been taken out in the car almost every day in order to allow him to meet other people and their pet dogs and when at last the big day arrived for him to be taken out for a walk on a collar and lead for the first time it was almost an anti-climax. Rio strolled confidently along the street and was gently stroked and encouraged whenever he was walking in the most comfortable position for his owners and not pulling on the lead. Whenever he put pressure on the lead, his owner stood still, gave a sharp flick with the hand that was holding the lead and praised and encouraged his dog when he returned to the correct position. This flick on the lead was designed to remind Rio that his owner had stopped all forward movement. Because he was only stroked and spoken to gently when he was in the right position when walking, the amount of pressure that Rio exerted on the lead rapidly decreased until he walked without pulling at all.

On many of the walks Rio would often see other dogs out with their owners, but, because he preferred games that his owners played with him and their company, he was very easy to control. This did not mean that he would not interact with other dogs. He was quite happy to socialize with any dog that came over to him but other dogs were certainly not an obsession in his life.

By the time that Rio reached the age of ten months his owners were able to list the following positive things they felt their dog had learned from them:

1. He was easy to control when around other dogs and would return readily when called, even when playing with his friend Ben in the local park.
2. Walking him was very easy as he did not exert any pressure on the lead and would readily change direction when asked to do so. This was extremely important, owing to his size and strength.
3. Not only did Rio fully accept being restrained but he could also be groomed and handled easily. This was most noticeable when he was taken to the vet to have an ear infection treated. Not only did he allow the vet to examine him but he also allowed his owners to administer the ear drops that were prescribed to treat the condition.
4. Rio had learned the rules of the household well and even in the unlikely event of his owners having to chastise him verbally for some minor misdemeanour, he accepted this 'punishment' without question.
5. Rio's behaviour towards guests was extremely good considering the fact that he did so adore people. After being allowed to go up and greet visitors he was then told to settle down and would readily do so although he still had to be watched to make sure that he continued to comply with these instructions as his enthusiasm and youthful years occasionally got the better of him.

Seeing an advertisement at the local library for a dog training club, Rio's owners decided to go along in order to learn some training techniques that would enable them to get the maximum pleasure out of their pet dog.

At the dog training class

At the first night of the new course at the local dog training club fifteen dogs and handlers assembled in the hall and the instructor commented on how unusual it was to see two blue Great Danes. The rest of the pupils were a variety of different breeds such as Collies, Shepherds, a Boxer, Golden Retriever, Poodle and Labrador.

Before the class got under way the instructor suspected that Roger would cause trouble because he had spent the first ten minutes, whilst everyone was arriving, taking the lead in his mouth and tugging against his embarrassed owner. The arrival of other dogs resulted in him barking uncontrollably whilst his now agitated owner looked on, helpless to prevent the behaviour. When the instructor asked the owners to walk their dogs around the floor to familiarize them with their surroundings, Roger promptly dragged his owner to the dog in front and proceeded to try and pin it to the floor with his front legs. When the instructor tried to intervene by taking the lead and to give him a hard check on the collar, Roger spun around and growled menacingly. A second check was applied which was designed to stop the growling. This time Roger took a pace forward and his owners could tell that he meant business for they had seen him in this position before with them. The instructor gingerly handed the lead back which helped to defuse the situation but Roger had won the day and would now begin to go from bad to worse at this class because he had found that he was physically stronger than anyone else there, a fact he had already established at home with his owners. Roger's owners spent the entire first night, just as they were destined to spend many more nights at this class, trying to keep him under some sort of control and to reduce the nuisance that he was causing to other pupils.

Rio by contrast arrived at the class and sat patiently by his owner's side waiting for the proceedings to begin. He was of course interested in the variety of dogs and handlers that were coming through the door but was quite happy to remain next to his owner. Rio seemed quite unconcerned when a fracas developed between two dogs and looked to his owner for comfort and guidance. Rio's owner was able to devote most of the time available to listening to and practising the training techniques the instructor described and demonstrated.

When Roger's and Rio's owners got together during the tea interval they were amazed to find that they were litter brothers. Roger's owners, after seeing Rio, immediately thought that they had been sold a particularly naughty puppy out of that litter and tried to explain that they had come along to a dog training class in order to gain a measure of control over their wayward pet. Rio's owners knew that they had put a great deal of time and effort into educating their pet dog and after seeing Roger they knew that it had been worthwhile.

Achieving control of a large breed

The first thing to bear in mind is that the earlier you start the better it will be for both you and your dog. The ideal age to begin control training is before your dog is ten weeks old. This is most important for the larger breeds because they grow at such a pace that it is undesirable to wait until the dog has reached a size where physical restraint becomes difficult. It is much easier to teach restraint and handling whilst the dog is still small and comparatively weak. Waiting until the dog grows in strength and confidence is counter-productive in the case of a larger breed.

There are two techniques which should be practised simultaneously in order to teach your young dog how to accept your authority without question. I refer to these two techniques as *physical restraint and handling* and *control games and exercises* and their purpose is to place the owner gently but firmly in control of the dog.

Physical restraint and handling exercises

Obviously the correct time to teach physical restraint and handling exercises is before your dog is physically strong enough to compete against you and so the sooner you do this the easier you will find it.

Begin by putting a snug-fitting collar on your dog, leather or soft nylon being ideal. Avoid using chain collars, half-check collars or choke chains. Now either have someone hold on to the lead, keeping it short so that movement is restricted, or fasten the free end to a solid item of furniture. Using one hand, gently lift your dog into the stand position, supporting his weight by placing your arm just in front of his hind legs and gently but firmly gripping the top of his leg with your hand. Talk to him in a reassuring manner and begin brushing down the outside of the leg nearest you. If your dog starts to struggle simply hold on until he stops, at which point you can begin brushing him again. It is important that you do not raise your voice or try to threaten your dog if he starts to struggle as that will only teach him to fight harder. The idea is to teach him as gently and as firmly as possible that it is pointless to struggle as this behaviour will not cause you to stop the grooming process but will, in fact, prolong the proceedings. If your dog attempts to turn and bite at your hand or the brush as you groom him then you should simply ease him away from the point where the lead is attached and take the weight off his hind legs as you do so. This now puts you in complete control as your dog is not only in a position where he will find it impossible to bite you but he will also feel vulnerable at being immobilized in this way. As soon as the struggling stops slowly allow him to take his body weight back on to his legs and restart the grooming process.

A tactic that some dogs use to prevent their owners restraining and handling them is to scream as if being seriously injured or else try to throw themselves on to their backs in a temper tantrum. If this happens then firmly hold the dog and continue grooming as if nothing were happening. When your dog has stood perfectly still to have one rear leg groomed turn him around, using the point to which he is fastened as a pivot, and groom the other leg. If, when you stop the grooming session, you follow it with something pleasant like a game with a favourite toy or a meal your dog will soon begin not only to accept these grooming and handling sessions but he should start to look forward to them as an exciting event. When your dog fully accepts his hind legs being groomed and handled you can then progress to brushing in between his legs and then under his tail, finally working over every inch of his body. Do not be tempted to rush these sessions, it is better that your dog learns his lessons thoroughly at each stage.

As your confidence builds and your dog accepts being handled more and more you should allow other members of your family to take a turn at grooming sessions. It is important that children only groom a dog when the adults are satisfied that the dog accepts this operation and even then it should only be carried out under adult supervision.

It is also a good idea to teach your dog to accept being restrained on a lead by simply attaching a lead to his collar and teaching him to spend ten minutes, twice each day, with the lead fastened to any convenient point in the room where you are sitting. These sessions must of course be supervised so that the dog does not become tangled in the lead but learns to accept being fastened, away from its owner, for a short period without objecting. If you ignore your dog if he makes a noise or struggles to free himself, and talk to him in a pleasant, reassuring tone when he is behaving in a calm and settled manner, he will soon adopt the calm behaviour that you are rewarding.

Control games and exercises

Most owners who have control problems with the larger breeds of dog actually lose control because they play inappropriate games and allow the dog to dictate the rules under which these games are played. It therefore follows that if you play the right sort of games and set all of the rules yourself then you will stay in control of your dog.

Games you can play

All tugging games are acceptable providing you can:

(a) use a toy that you can easily get back off your dog whenever you want;
(b) place the toy on your knee at the end of the game and prevent your dog touching it by telling him to 'leave';
(c) call a 'time out' if he becomes overexcited and begins to growl; and
(d) get your dog to bring the toy to you after you have thrown it for him.

All chasing games with a toy are acceptable providing you can:

(a) get him to give up the toy easily on request;
(b) call him back when he is running after the toy that you have thrown without allowing him to touch it;
(c) make him stay whilst you throw the toy and occasionally go and pick it up yourself, leaving him to wait for you to return with it; and
(d) use a toy that is completely safe for him to play with.

Games to avoid playing with your dog

Avoid games that:

(a) get your dog overexcited to the point of becoming out of control;
(b) involve using yourself as a toy;
(c) involve encouraging your dog to put his teeth on your body or clothing;
(d) involve wrestling or allowing your dog to pit his strength against you and win the game; and
(e) encourage your dog to exhibit behaviour that will get him (and you) into trouble when he gets bigger.

If you ensure that all the family members fully understand how to play games with your dog and make sure that they abide by the rules of the game you should all end up in total control of your dog as he grows into an adult.

Remember that games, more than anything else, teach skills and techniques that will be of use later in life. It therefore makes sense that your dog learns how to develop skills that will be of benefit to you and your family as he matures. Control the games, control the dog!

6 Separation Anxiety
Margaret Goddard

This chapter will examine one of the commonest behavioural problems, that of separation anxiety, and how it affects dogs. For reasons that will be discussed later, the condition affects cats much less. 'Separation anxiety' is a state of anxiety that can arise when a dog is left by or separated from its owner.

Anxiety is a state that is common to humans and dogs. In humans it manifests itself in feelings of uneasiness, apprehension and foreboding, which seems to be true of the anxious dog as well. When it is present, definite recognizable physiological changes occur that originate in the central nervous system, such as increased heart rate, tremors, restlessness and gastro-intestinal symptoms, to mention but a few. At a certain level of anxiety functioning can be quite normal and indeed it can serve to improve performance, but at higher, more intense levels it will disrupt normal activities and cause distress instead. Certain anxious individuals have a raised level of arousal in the central nervous system which makes them react more excitedly and adapt less well to certain events. The distressing symptoms, being unpleasant in themselves, will often reinforce the anxiety.

Separation anxiety can show itself in dogs in many ways that occur either singly or as a combination. The common signs are vocalization, either barking or whining, urination and/or defecation, and destruction. There may be depression, which shows itself in the dog not wanting to eat or drink, or appearing very withdrawn when an owner is about to leave. Loss of appetite or lethargy can occur if an owner is away for a long period, a holiday for instance. The dog may become hyperactive when an owner tries to leave and mouth, nip or growl. These latter behaviours are often displayed by a dog that has an obvious dominant aggressive nature at other times. More rarely one may see excessive salivation, vomiting, excessive licking or hair chewing as alternative displacement activities when the dog is left in a state of anxiety.

Dogs that suffer from separation anxiety are usually obedient and well-behaved pets in normal circumstances. In fact, these dogs are often extra loving and attentive towards their owners. Likewise, the owners are often very deeply attached to their pets and therefore find it very hard to comprehend why their pets destroy their owner's hard-earned possessions when they are out or urinate or defecate all over the floor. They say things like, 'I give my dog so much affection and attention yet it "pays me back" for being away'; 'My dog does these things "in spite" because I leave him alone.' These are commonly held beliefs that express the mixture of guilt and anger that owners often feel. There is guilt because they wonder if they are at all right to leave, and anger because their pet has dared to destroy belongings or cause other problems whilst they are away.

The first thing we must realize is that there is no 'spite' or 'paying back' involved. Dogs do not understand these concepts and are therefore incapable of expressing them.

They live for the present and do not either dwell on the past or think about the future. What they do is to learn responses to problems and situations and repeat them. Thus, some dogs have learnt not to be able to cope when left alone and so become anxious. In some instances, this can develop into a separation anxiety problem. It is clear, therefore, that the dog's past may influence its present way of life. To build up a picture of how some of the behaviours we are concerned with may have come about we need to examine the beginning of the learning processes of puppies and to follow their development.

Dogs are social animals and live in packs that have a definite hierarchical structure. When a puppy leaves its 'pack', i.e. its mother and litter mates, it joins another pack, the human family. Social animals have a drive for social contract. Thus, when separated from their mother or parents, young animals will try to maintain contact through a variety of instinctive behaviours. Birds cheep and chirp, babies cry, puppies whine, lambs bleat, etc. When the young are re-united with the parent, this activity ceases. As the animals develop, they become able to tolerate a certain amount of isolation. When a puppy is brought home by its new owner, it will readily transfer its social attachment, but there can be a degree of separation behaviour until it learns to tolerate some degree of social isolation.

Up to the age of three weeks, the puppy's needs are very basic regarding sleep, food, warmth and stimulation to urinate and defecate. The pup has simple reflex behaviour to deal with these needs and likewise the bitch has instinctive behaviour which helps to satisfy them. Anything which disrupts this bitch–pup relationship can disrupt the future learning and character of the pup.

A bitch's nervousness can be passed on to her pups. If a pet bitch is very attached to her owners, forcing her to be with her pups when she would rather be with them will cause her to be restless and agitated and to unsettle the pups. Owners should recognize this possibility and allow the bitch to choose where she and her pups would rather be. In so doing, she will remain calm and settled. Again, bitches kept isolated and used as breeding machines, which sadly occurs on puppy farms, and even in certain kennels, are not likely to be competent in coping with life's many changing situations. They will often produce puppies that will show themselves equally incompetent in dealing with situations later in their lives. Paradoxically, some of these puppies become over-dependent on their future owners and thus are unable to cope on their own.

As David Appleby explains in his chapter, early learning and socialization are all-important in producing a puppy well adapted to the situations it may come across later in its life. However, genetics also plays a considerable part. Even if a pup has been reared in the right environment and given correct training and handling once acquired, the animal's basic genetic make-up will still influence its ability to cope, learn and adapt in future. It must be remembered that some dogs, no matter how perfect their handling, will have characters ranging from the apprehensive to the extrovert because of their inherited genetic make-up.

Certain breeds are known to be over-representative when it comes to separation anxiety. German Shepherd Dogs, Boxers and Labrador Retrievers are amongst some of the breeds regularly seen with this problem. They are notorious for being chewers when they are young. It is a natural instinct for puppies to chew when teething but some enjoy chewing and mouthing throughout their lives. Unfortunately, if the pup is of that disposition, this trait may develop into a problem if the pup is left alone a lot. Nervous bitches may pass this quality on to their pups.

We have evolved to live in social groups and in close proximity with other animals from which we derive mutual benefits. Dogs are ideally suited to being companion animals because of the ease with which they form strong attachments to people. Dogs are individuals and will react as circumstances and their basic nature dictate such that one cannot tell why some dogs suffer from separation anxiety and not others.

The emotions owners feel when a dog begins to exhibit behaviour associated with separation anxiety are very severe, yet these can compound and reinforce the dog's behaviour. If a dog becomes attached to us we feel satisfaction at having a pet that adores us, it makes us feel good. We feel wanted, loved and needed; and, conversely, we feel guilt when we have to leave our pet and are unhappy. If a dog then develops a separation anxiety problem an owner may feel more guilty, or even angry at being let down by their pet. They may react by showing more love and affection when they are present, not realizing that this will increase the inability of the dog to cope on its own when left. Owners may react by shouting at their pet, beating it or even rubbing its nose in a foul-smelling pile of faeces. This again compounds the problem.

On the dog's side there will also be a tremendous conflict of feelings. It may be overly attached to its owners and unable to cope on its own so it is excited and eager to be reunited with them. But what happens when they return? It receives punishment in some form. This soon causes immense apprehension and fear, which gives rise to further anxiety. The stress of conflicting feelings compounds the anxiety a dog feels when left.

A dog is a master at reading body language and is sensitive to its owner's moods. A dog learns quickly to look and act guiltily by the demeanour of its owners. It has absolutely no conception of having done wrong. It simply reacts to the owner's state when he or she enters the room and if it learns that an owner's return is associated with punishment it will react accordingly. If it is surrounded by excrement or debris it will also act guiltily. This is because it has learnt that the presence of these means punishment, yet it cannot associate causing the mess with punishment.

For an owner to realize that these feelings are normal will often relieve the pressures of conflict when their pet does wrong. This may be sufficient to reduce the anxiety felt by the dog when it is left such that it no longer engages in problem behaviour.

It has been explained how the seeds of a separation anxiety may be sewn in early puppyhood, but other circumstances can arise which make some dogs unable to cope with separation. Family rows and upsets can lead to increased stress in the anxious dog which may lead to an inability to cope when alone. Situations which involve dramatic changes to the daily routine can predispose dogs to develop a separation anxiety, as when marital break-ups result in a dog being sent to a rescue centre. If the dog has a separation anxiety when it is rehomed it can very soon find itself back at the centre because of its behavioural problems.

If an owner is at home for a long time, on holiday or unemployed, the dog may forget how to cope on its own when the owner returns to work. A part-time worker who changes to full-time employment may find the dog cannot cope with the longer separation. Likewise, dogs that are infrequently or never left alone may never learn to cope.

Some dogs are able to cope with the first departure of the day, but if the owner returns and later makes a second departure the dog may no longer tolerate the isolation.

External events such as the doorbell ringing or the postman arriving can cause panic or fear in some dogs. If the owner is not present to comfort and allay the fears of the dog, these events may trigger off anxiety. Other dogs become excited at hearing external noises and so destruction or vocalization may result.

The dog that has always been in the company of another pet may not be able to cope on its own after the death of that pet.

Having understood the problem, the owner then has to decide if he or she is willing to adopt the measures that will lessen the anxiety felt by the dog when it is left and thereby extinguish the unwanted behaviour. Behavioural psychologists describe anxiety as a learnt response to various stimuli that eventually becomes a deeply conditioned habit that cannot be controlled and is brought into play by the slightest stimulation. As it is a learnt response, it can be 'cured' by teaching a different response. In its place we develop a pleasant response, thereby extinguishing the previous bad habits. However, this new response is only brought about by making the dog less dependent on the owner, and it is the price the owner has to be willing to pay for the cure.

Some owners are just not able to pay it, and they can be those who care too much as well as those who care too little for their dog. For some, distancing themselves from their pet in order to reduce the level of dependency is not a practical proposition because the dependence *is* the point of the relationship for them. At the other end of the scale, if the owner has never formed a strong bond with the dog it is not likely that they will be willing to commit themselves to the work involved.

Those who cannot bear to change their relationship with their pet, for whatever reason, are faced with two choices: one is to accept the situation and take avoiding action, the other is to get rid of the pet by rehoming or putting it to sleep.

Avoiding action may mean never leaving the dog, which takes many different forms. Dog sitters or friends are often employed to be with the dog when the owners have to be away. Some owners will always take the dog with them wherever they go, even to work. (Many dogs will be perfectly content in a car whereas they cannot be left alone in the house.) Other owners never visit people or go out if the dog cannot be taken along, which can seriously affect their social activities. Some kennel the dog outside to prevent destruction or urination and defecation becoming a serious problem. The expense needed to set up a safe construction of this type is considerable so other owners settle for the option of having one room in the house just for the dog. Indoor kennels are another solution that people try.

Everything considered, the expense and trouble involved in dealing with a separation anxiety problem in this way can be considerable. In my opinion, the best way of dealing with the problem is to solve it.

Two separate lines of action are undertaken in treating separation anxiety problems. The first has already been mentioned, that the relationship between owner and pet has to be altered to decrease the dependency of the pet on its owner. The other is to desensitize the dog to being left by its owner. The former can be carried out straightaway but the latter is best left to when the owner has the time to spare and the effects of the former changes have become apparent. A holiday period is a suitable time to start this. The single most important fact for an owner to remember is never to punish their dog on returning to the house, no matter what has happened.

Usually, pets with separation anxiety problems will follow their owners around the house – even to the extent of lying outside the bathroom when the owner is inside. They will lay at their owner's feet or climb on to their laps, if small enough, to enable close contact at all times. At first, the dog must be taught to be away from its owner. Ideally a special mat or place should be allocated in different rooms in the house. The dog must be taught to lie quietly in this place when asked to. Titbits can be used as rewards. If necessary, the dog's daily food ration can be reduced and replaced with titbits to hasten its compliance.

If the dog prefers to lie in doorways or passageways or at the foot of the stairs it should be made to sleep elsewhere. These places are preferred by dominant dogs, by which I do not mean aggressive ones, but those who do what they want when they like. Some dogs with separation anxiety problems are always dictating to their owners what to do, e.g. instigating games, asking for walks, asking for affection. These are the more dominant-natured dogs. It is the owners who must instigate activities and they must learn to ignore the dog's commands.

A babies' stair-gate in the doorway will prevent a dog following its owner around the house, yet allow visual and auditory contact. Dogs with separation anxiety problems often sleep on the owner's bed or are allowed to sit on furniture beside their owners. This must not be allowed and again can be prevented by using the stair-gate. Ideally once the dog has learnt to rest away from its owner the dividing door should be gradually closed, but only if no signs of anxiety are seen.

A general, simple, daily obedience session is useful. This helps the dog to learn a calmer way to behave which then becomes a new, learnt behaviour. Asking the dog to sit or lie down before putting the lead on or before feeding makes the dog more obedient to its owner and lessens the extent to which it lives its life the way it wants to. A dog that has learnt to sit–stay and down–stay is being treated like a dog and even the simple act of teaching these commands will separate it from its owner to some degree and help to reduce its anxiety.

In the early stages of this programme, the dog may often seem withdrawn or sulky. This is only a temporary phase. The dog is adjusting to its new way of life and to not being allowed to live life the way it wants to.

Titbits should be given only as rewards for good behaviour and not to satisfy the owner's need to give comfort and to please their pet. At first a dog will pester the owner for titbits and attention but soon, if the owners persist, this behaviour will stop. It is very difficult to deny the appealing looks of a beloved pet when it asks for titbits, walks, affection or attention, but it must be done.

Some dogs with separation anxiety problems benefit from having a den. This usually means an indoor kennel. With proper introduction and learning it can be a useful adjunct to therapy. To understand why, we need to look again at the dog's learning patterns. When the pup is born, the mother will provide food, shelter, warmth and security in a quiet place or 'den'. As a pup develops and begins to explore further from its nest, the importance of this den is lessened. Some dogs with separation anxiety problems feel calmer and happier if secured in a den and therefore do not engage in abnormal behaviour.

However, for this to occur, the den or indoor kennel must be a fun place for the day and never be associated with punishment. To begin with the kennel or cage must become part of the dog's daily life. The dog should be encouraged to sleep in it and meals should be given in there. Affection and attention should be given when the dog is in the kennel. Instigation of games or walks should begin from the kennel. Only when the dog is happy to go into its kennel and stay there should the door be shut, and in the beginning this should be for short periods only.

Some dogs will not accept close confinement. It simply exacerbates their problems or causes others. Dogs that do not tolerate confinement to a kennel or even to one room in the house may be perfectly behaved if they have the full run of the house, as they do when the owners are at home. They then have the opportunity to move around and choose where they want to be. It does, however, take a brave owner to try this approach.

Some people mistakenly believe that getting another dog as a companion will alleviate their first dog's loneliness when they are out. This may work in the case of a dog which has suddenly lost a companion but rarely works in other cases, the reason being that the new dog's company is no substitute for the close attachment the dog has with its owner. Indeed, the new dog may be taught the same bad habits thus doubling the problem.

It is possible that these desensitization changes will suffice to calm a dog to the point that it will not suffer from a separation anxiety problem any more. If not, the next step is for the owner to de-emphasize the importance of their departure and return.

Many an owner compounds the anxiety of a dog when leaving by prolonging the goodbye, by cuddling and sweet talking and perhaps telling the dog, 'Mummy is only going to be away for a short while and will definitely be back in an hour'. Dogs have no concept of time. They feel the same whether they are left alone for twenty minutes or two hours. For the same reason, dogs will give the same degree of welcome to a returning owner whether it be after twenty minutes or two hours. Dogs cannot understand English, no matter how convinced an owner might be that they can. They learn to associate such words as 'sit' or 'walkies' with certain situations and tones of the owner's voice. Saying 'You horrid, nasty, despicable dog' in a loving, sweet-natured voice will elicit a tail wag and a loving, pleasing manner from a dog, but say it in the appropriate tone and the dog will grovel with ears flattened looking thoroughly miserable and upset. Giving extreme affection and love just prior to departure heightens the anxiety in a dog. The dog wants this social interaction to continue. It is instinctively programmed to want this. A sudden stop to this interaction leads to acute stress and anxiety.

First and foremost, behave calmly before leaving a dog. The dog should be 'sent to Coventry' for an hour before the owner leaves so that it is not a major crisis when he or she departs. In addition the dog will be less excited if it has been well exercised and fed prior to the owner's departure.

The owners must also change how they behave on their return, particularly in respect of 'punishments'. As I have mentioned before, punishment has no place in the treatment of separation anxieties. A dog that has learnt to behave in a guilty, fearful manner because it is usually punished, or the owner is usually angry, will soon return to its normal self once punishment ceases. The typical anxious dog often seems to go 'over the top' when greeting its owners on their return. Frantic whining, barking, jumping up, tail chasing, panting and fetching a toy are all commonly seen in these dogs. Arrive in a calm manner and matter of factly greet the dog, ignoring its over-excited manner. It helps the situation if the owner walks in backwards and only turns to face the dog once it has calmed down, thus rewarding its calm behaviour with the visual and sensual contact it craves. If the owner shouts or gets excited it just rewards and encourages the dog's bad behaviour and guarantees its repetition. Periods of social interaction, such as walks, grooming, feeding or playing games, should also be rescheduled to coincide with the return of the owner.

Tape recorders can be used to find out what happens when the owner is away and as an aid in treatment. A tape recording of what happens in the house when the dog is left shows how long it takes for the problem to begin. If a dog has intermittent problems it may be found that the telephone or doorbell ringing or the postman arriving are the contributory factors. These situations can either then be avoided or treated as one would for most phobias. A tape recording of normal household noises, or a radio left playing, will help a dog feel it is not alone.

Old garments left with the dog may create in it a feeling of being close to the owner. In some cases, dogs allowed into the bedroom when left feel comforted whereas they become distressed in the kitchen.

Chewing and destruction seem impossible to treat as they occur in the owner's absence, but there are strategies that can help. Dogs often chew doorways, scratch at the walls or destroy the floor next to the door through which the owner departed, and are the dog's way of trying to maintain the contact with its owner that it desires. Often chewing will act as a pacifier to the dog. If cardboard boxes and pieces of wood are left, dogs with a destructive problem may choose to chew these in place of the furniture or door frames and this may be considered acceptable.

Chewing and destruction can sometimes be self-rewarding to the dog. A classic example of this is when cushions are destroyed, the eruption of stuffing from a cushion often causing further excitement and so acting as a reward. Opening cupboards and finding food is also self-rewarding. If a specific article is chewed, denying access to it may help. Booby traps, such as a pile of empty cans, or spraying the object with some bitter substance may prevent destruction by punishing the dog immediately chewing begins. The punishment must be associated with the act of destruction. Punishment by itself will never decrease anxiety. It only helps to suppress or reduce a behaviour. In some cases it can actually increase anxiety as the dog will become more fearful and want more security and contact with its owner. Hence it may change to different behavioural problems or learn to delay its aberrant behaviour.

This is precisely the reason why special collars designed to stop dogs from barking may not work, and in fact may create further, far more damaging problems. A dog that barks or whines when left is only trying to maintain contact with its owner, to call him back. It is just like a pup calling for its mother. If this is prevented its anxiety may be heightened thus leading to further problems.

Some dogs will reach a peak of destruction and then cease, whereas some dogs are capable of barking or howling all day. The only answer to all this is to desensitize the dog to the departure of its owner, which must be undertaken calmly and in the knowledge that it is a time-consuming business. It is also repetitive, boring work – but ultimately very rewarding.

As explained, dogs are masters at reading our body language, actions and the tones in our voices. Usually when people prepare to leave the house there is excitement and action occurring. Voices are raised, people move from one room to another. They visit the bathroom, pick up coats, put on shoes, pick up keys, put hands on door handles – all steps that enable a dog to recognize a departure is imminent and that arouses and excites it.

To desensitize dogs to being left, the owner must break down their actions in detail. As much forward preparation as possible should be undertaken, such as putting articles in the car long before a departure is imminent and getting all belongings together in one room.

An owner should take these preparations only to the last point where the dog will remain calm and non-anxious in its bed or den. Repeat this many times, and always reward the dog for its good behaviour. If the dog becomes anxious, the owner is trying to progress too fast or has not recognized the point to stop. Gradually the dog learns to remain calm. Once the owner can pass through the door and close it, the intervals of departure should be varied. Departure times of one, three, five, two, seven, four, one, two, ten, six and so on minutes should be tried. If the dog becomes at all anxious, the owner must ensure it returns to a non-anxious state before starting again.

The importance of not hurrying this process cannot be over-emphasized. The owner must work at the dog's learning pace. Again, the use of a tape recorder before embarking on the desensitization programme will tell how long a departure a dog can tolerate before becoming anxious and give a basis on which to work.

At the end of this process most dogs will be 'cured' and will remain so. However, if an owner's treatment of their pet reverts to the over-attentive, clinging manner of before, the dog's behaviour will also revert. Moreover, if all the new changes are recent and not yet reinforced and imprinted the dog may well be worse than it was before. Consider the true case of a Staffordshire Bull Terrier, six months out of quarantine. The owners suddenly changed jobs. The dog could not cope with the lengthened separation and began destroying the room it was left in, despite having another dog as a companion. Once all the measures described were undertaken, very little further damage was done and the dog returned to its normal calm state. A few weeks later relatives from abroad arrived. Sadly, despite being told not to 'spoil' the dogs, they proceeded to give undivided attention, treats and titbits to them. When the visitors departed and the owners returned to work, the dog reacted worse than before, destroying an entire wall in the house. The owners' breaking point came when the dog destroyed the interior of their car having previously been content to stay quietly in it. With great sadness they decided they had to part with their beloved dog. However, fortune smiled and they rehomed it with a lady whose situation was such that the dog would never be left alone. This salutary tale shows that one should not underestimate the complexities of a dog's behaviour.

Drug therapy has been tried as a method of treating separation anxiety and it has been found that, while drugs can have a place in the correction programme, in some cases they are of no use on their own.

Beta-blockers have been used to try to prevent the anxiety spiral developing. However, the dose required to alleviate the anxiety can cause very serious side effects such as collapse.

Herbal and homoeopathic remedies have in some cases contributed to the treatment of separation anxieties.

Various sedatives have been tried. Occasionally, these can cause hyperactivity, rather than sedation. With some sedatives the dog's sensitivity to sound can increase. They have the unfortunate effect of inhibiting the dog from learning at the dose needed to reduce its level of arousal. When the drugs are reduced or stopped the original problems begin again. In addition, these drugs soon require higher and higher dosages to maintain the same level of sedation and so cannot be used for a long time.

Certain hormonal drugs can help by influencing the central nervous system, making the dog calmer and less anxious. However, one side effect can be an increased appetite. This may drive a dog to destroy cupboards to get to food even if they refrain from the previous separation problem. In dogs such as Labrador Retrievers hunger is a very potent force.

As with all behavioural problems, the owner must check that there is no underlying medical reason for a dog engaging in a separation anxiety problem, especially if it is only a recent phenomenon.

In this respect a commonly seen problem is that of a dog which is well behaved when left during the day but starting to defecate when left overnight. Usually this follows a period of change such as the dog living in kennels or having been away on holiday with the owners. Separation during the night suddenly seems to be beyond the dog's tolerance level. There then needs to be an investigation to determine if this loss of normal toileting behaviour is due to the dog being kennelled for long periods, true separation anxiety or in fact a condition resulting in colitis which appropriate medication and dietary change will cure.

Another case to mention is that of a Border Collie that was of the hyperactive, sensitive type. Over a period of some months, her behaviour changed. She readily panicked when out of sight of her owners on a walk and began to urinate and defecate when left alone. She became more nervous of some situations yet apathetic to others. In total, her character was becoming unpredictable and a worry to her owners. After investigations, a brain tumour was diagnosed as being the cause of the problem.

As mentioned, in drug treatment hormones can play a part in therapy. An eight-year-old Bassett Hound started to become more anxious when left and showed signs of distress prior to the owner's departing, although no real problems occurred when it was left. After disappearing to hunt on a walk, the dog got into a state of great distress when it realized it had lost its owner, although they were quickly reunited. The following day, when the owner came to leave, the dog worked herself up into such a state of anxiety that she had an epileptiform-type seizure lasting several minutes. By detailed questioning of the dog's previous history, this problem was traced back to it having had an ovariohysterectomy for a pyometra. This classically arises when a bitch is going through a period of great hormone change. A course of the appropriate hormone treatment settled the dog back into her normal behaviour patterns.

The problems with cats will be covered in more detail in Peter Neville's chapter, but in my experience recognizable separation anxiety problems are rarely seen in cats, and if they do arise they are not as harmful to the cat–owner relationship as they are with dogs. When the owner is away from home, the cat is likely to become quiet rather than to indulge in displacement behaviour. On the owner's return, the cat will follow its owner around and ask for attention. If stressed some cats may indulge in excessive self-grooming when left, which can result in hair loss or alopecia. There can be many causes of the stress that can lead a cat to overgroom. Normal grooming, it is believed, in addition to cleaning the coat and ridding it of loose hair and dust, gives the cat a feeling of calm and well-being.

Until recently many cases of alopecia were described as being caused by hormonal problems. One reason was because the hair loss was bilateral or symmetrical and characteristic of the well-recognized hormonal problems in dogs. The other reason was that a very successful drug used in 'curing' the problem was a potent hormone. Now it is recognized that alopecia without damage to the skin is usually a psychological problem. The drugs used to cure the problem have potent effects on the central nervous system and produce a sedative, calming effect, thereby alleviating the stress and 'solving' the problem. Unfortunately, most owners seem happier to give their cats regular courses of these drugs rather than bother to try and resolve the reason for the stress, which of course involves considerable effort.

In rare instances, cats go beyond overgrooming and self-mutilate. This is seen as thickened areas of inflammatory tissue on various parts of the body but particularly on the legs or ventral abdomen. Stress from various reasons may be the cause, but other medical problems can give rise to this condition.

When some cats are put into a cattery they seem to suffer from separation anxiety. They may eat very little or else refuse all foods. They may sit hunched up and miserable. Some stop grooming themselves. Some will even lose their toileting habits. Owners like to believe that the cat is pining for them. Unfortunately, the truth is more likely to be that the cat is miserable because it has lost its familiar surroundings and smells. Such cats may settle if toys and belongings from home are placed in the cattery with them. If the cat still will not settle, it is kinder for it to be at home with a neighbour calling in to feed it on the next occasion it has to be left.

These days many previously inoperable conditions can be successfully treated, such as broken pelvises or fractured jaws. Total dependency on its owners for weeks or even months for food, warmth, grooming even toileting can lead a cat to expect this treatment and to become anxious when the owner departs. Usually, the cat's natural habits take over when it returns to full health and the dependency declines without the owner having to do anything to correct it.

A similar type of dependency can arise in hand-reared kittens or those that have left their mothers before learning more social skills. The kitten transfers its attachment from its mother to the new owner. Some will regularly suck on their owner's clothes, some will be slow to develop toileting habits, some will be slow to learn how to groom themselves and some will not let the owner out of their sight. Generally, with time the cat's strong natural instincts will take over and so it will not be a problem to the owner.

It is always more difficult to establish the cause of a problem in cats than dogs as stress can occur due to a multitude of reasons.

Lastly, how do we prevent a new pup from becoming too attached to us?

Ideally, before a new pup is brought home, the new owner should take one of their blankets to the breeder and place it with the litter, then the pup already has a familiar smell surrounding it when it arrives in its new home. At first, the same food should be given and the same feeding regime adhered to as the breeder used.

The first night away from its litter mates can be traumatic. A warm, comfortable bed is a must. Pups will often find comfort in snuggling up to a soft toy. The old idea of a large alarm clock ticking to simulate the mother's heart beat is somewhat outdated, but certainly quiet background noise prevents the feeling of total isolation. A warm, milky feed before being put to bed will create a feeling of comfort and drowsiness.

A pup will either successfully sleep through the first night or cry and whine to some extent in an effort to attract social contact. The owner may decide to be firm and not respond to the pup's cries in the hope that the crying will diminish with time. If crying persists a calm but cool check on the pup with no fussing may be enough to reassure it that its new owner will return. The alternative is to take the pup up to the bedroom, but with the determination that once it has settled it will be gradually moved out of the room to its designated place of rest over the succeeding nights.

Pups come from a situation where the 'den' or 'nest' was a safe, secure, comforting place to be and will usually readily adapt to an indoor pen or kennel. This is ideal for helping a pup become accustomed to being left and will prevent inappropriate chewing.

A new owner who understands the causes of separation anxiety can begin to treat the dog in an appropriate way to avoid over-dependency later in its life.

It is imperative that a new pup is never punished for toileting problems when young. Up to six months of age, pups cannot be relied on to go through the night without an accident. To punish for a problem that is no fault of the pup will certainly cause severe anxiety in a pup desperate to greet its owner in the morning and when left for a long period.

It is unreasonable to expect a pup to be on its own for long periods without becoming bored. Chewing could well ensue and become a definite habit, not just a passing phase.

Rescue dogs can be an unknown quantity. The real reasons why dogs are in rescue centres are rarely known. Much the same methods apply to older dogs as for pups. However, older dogs can take longer to learn new ways and patience is needed. Until its behaviour is known, it is a good idea not to leave the dog on its own for too long. Older dogs often travel successfully in a car, unlike a new puppy.

New owners should not allow themselves to feel they have to give 'extra love and attention' in order to make up for problems in the dog's past, as this can quickly produce a state of over-dependency in the dog.

In conclusion, separation anxiety problems can be resolved but as always it is up to the dedication, patience and tolerance of the owner. There are successful ways, but they are not easy.

7 Phobias

Robin Walker

The genesis of phobias

The genesis of a phobia is often a very dramatic or powerfully stimulating event which causes a burst of fear in the victim. The lesson to be drawn from the event is fixed at once and forever with the aid of the adrenaline the fear stimulated. Evolutionary encounters with snakes, polar bears or tigers do not permit leisurely repetitions of the lessons of experience.

It is now scientifically recognized that fearful or threatening events enhance learning ability, which is something schoolmasters and drill sergeants have known intuitively for generations. Inevitably such people tend to overplay their empirical methods and induce levels of fear that have the opposite effect, they wipe out the ability to learn! This complicates the learning of such difficult tasks as assembling weapons or working the winches on a yacht which require up to eleven repetitions to 'stick' or become properly learned in normal conditions.

Whether the response to threat is controlled or excessive depends on the circumstances of the initial encounter. In Wimborne in the autumn of 1960 the lid of a brass coal box fell on Ollie, a fifteen-week-old wire-haired Fox Terrier. He screamed with pain and fear and fled from the room. He turned in the doorway and barked hysterically at the coal box. As the lid was replaced the clatter frightened him again and the cycle of flight and barking was repeated. Sadly but alas almost inevitably Ollie's phobia became a joke or party piece. He was 'chased' with the lid or attracted into the room by the lid being rattled deliberately. In short his fear was reinforced for the entertainment of visitors. Ollie retained his fear and rage about the coal box lid until his death at fifteen years of age.

Another example is the case of Poppy, a six-month-old Jack Russell. During a Christmas party at the Royal Veterinary College field station, Poppy was pottering among the assorted adults and children quite happily. Suddenly she froze and shrieked with fear. She was staring at a magnificent blonde Rosebud doll with vivid blue eyes which stared unblinkingly back at her from an armchair. From that moment Poppy feared children. She had to be removed from the party because she began raising her hackles and barking at the children present. The only children Poppy met in the ensuing years were the urchins of Maidstone who delighted in provoking her to frenzied barking when they spotted her in the car. Such incidents as having sticks poked through the window gaps ensured that her phobia was lifelong.

In such cases, although the dog shows aggression, the overwhelming response is fear. If it is trapped with the source of fear a dog will make violent attempts to escape via windows (closed or open) or doors. Desperate attempts may be made to 'go to ground' or dig dens through carpets and floors or into wardrobes or furniture upholstery and beds. Sometimes the dog will collapse in flaccid paralysis of two or even all four limbs. Two unrelated Labradors developed phobias about walking on Marley Tile floors. If obliged or forced to cross this type of surface both dogs would show extreme fear, freeze and fall down with epileptiform seizures. One 4 November a six-year-old crossbreed with a phobia about people wearing blue jeans was trapped in the sitting room of its home with the 'Guy' made by the children for the next day's bonfire. Although the family knew of the phobia they forgot, dressed the effigy in an old pair of blue jeans and placed it on the sofa. The dog had a series of seizures for the greater part of twenty-four hours until the cause was burned. Every time he awoke there was the Guy! In the three years that I knew this dog there were no seizures from any other cause. The emotional stress of being afraid of the clipping parlour has induced partial and partial-complex seizures in Miniature Poodles. These cases seem to suggest that forcing the fearful dog into the situation it dreads can be highly counterproductive if not downright cruel.

Gemma: a case history

Gemma is a merle Border Collie bitch, eight and a half years of age, bred on a farm with working parents. Her history illustrates many of the features of the phobic case and the difficulty of unravelling the problem from the owners' account of it. From the outset Gemma's problems were perceived to be a form of 'panic'. She has a veterinary record:

13.8.88 Had a 'panic'. Destructive of furniture. Eyes bloodshot at the time. Four episodes in a month. Prescribed phenobarbitone tablets.
13.8.90 Two episodes of 'stumbling' and 'almost collapsing' on walks. No clinical findings. ?Jacksonian seizure.
16.8.90 Another 'turn'. Suddenly loses legs, topples over, 5-6

minutes. Panting and wandering afterwards. Won't sit. ?Partial-complex epilepsy.

2.11.91 Destroyed another door yesterday. Giving 60 mg pheno-barbitone when Mrs S goes to work. Full behavioural inquiry suggested.

13.3.92 Again reports resumption of 'panics'. Investigation strongly urged. Accepted.

With the aid of 1½ hours of taped recorded consultation, four months of observation and five telephone discussions, a picture of Gemma's affliction has emerged. It is possible to set out the facts and pose some questions:

Is she of a nervous breed? Were the parents nervous working dogs? To what extent was Gemma socialized up to 9 weeks when she was sold?

Gemma is nervous and somewhat aggressive in the consultation room and at the house. For the first four years of her life Gemma lived in a large house out in the country. Her first problem was a tendency to bolt home when taken for walks.

How effectively was Gemma socialized in her early weeks at her home? What frightened her on walks?

The episodes of flaccid paralysis of hind legs followed by anxious panting and pacing only occurred when the husband or son walked the dog – not with the wife.

What happened on the walks? Was there any noise? Was the dog trapped on the lead and unable to escape some form of terror?

The earlier 'destructive' behaviour in the house only occurred in the absence of the owners but by no means on every occasion. Attempts were made to enter wardrobes, beds were burrowed into and the bathroom mats were scuffed up into a pile.

Was Gemma frightened by something in the absence of the owners?

In the previous eighteen months the 'attacks' were getting worse. On returning to the house on the evening of 13 March the owners were greeted by a gust of 'smell' and a saliva-drenched dog in a state of shaking, panting terror which rushed past them and fled to the nearby home of their daughter. Here it was seen jumping up at the windows with bloodshot, 'bulging' eyes. Indoors the dining room carpet had been torn aside and the floorboards scratched and chewed.

Could this be separation anxiety? Why so infrequent? Has the dog ever panicked in the presence of the owners?

Yes! Gemma panics in the garden when the owners are present. She pants and paces back and forth and barks at the house windows as if she can see something. She does that indoors. By 25 March, when a house consultation is arranged, Gemma has had a full-scale panic attack with Mrs S present. While sitting quietly drinking coffee she noticed Gemma becoming distressed, panting, red-eyed, salivating.

She disappeared into the bedroom and attacked the wardrobe. The owner was unable to placate her and she continued wandering, hiding in corners and behind the settee.

Was there any noise of any type at any distance? What is the owner's reaction to the behaviour?

Mrs S's reaction has been one of extreme exasperation. Severe lectures have been given to the dog: 'Come here and look what you have done!'; 'I am at the end of my tether'; 'On Friday when I was telling her how bad she had been . . .' At other times the dog has received increased sympathy, soothing patting, more time spent walking and playing and been taken to work regularly in Mr S's van. 'Gemma *never* misbehaves in the van. Oh, she did dig up the van's contents once. Oh, and she did escape through the van window in the thunderstorm.'

What's that about thunder?

In the July thunderstorms Gemma exhibited her full repertoire of phobic responses. She also goes beserk when the letterbox rattles (come to think of it)!

Gemma seems to epitomize the case of noise phobia and now that her owners are beginning to listen carefully more may be discovered. She also shows how fear can become more general. When she is afraid in the garden she tries to escape into the house, and when she is afraid in the house she tries to escape into the garden. The progressive increase in fear suggests that Gemma is becoming more generally anxious in a non-specific way. This is perhaps chronic phobia or general nervousness, for example, fear of being outside, of people, of traffic or of fear itself (phobophobia). Perhaps she was becoming afraid of her owner's anger and exasperation, or perhaps her anxiety was heightened by all the sympathy and soothing which may have seemed to be a pack response to her fear. Her owner said consistently that she just seemed to want to escape. The final refuge for the desperate dog is the den, in which the natural or archetypal dog or wolf begins life. It is perhaps not surprising that dogs suffering from hypovolaemic shock due to gastro-enteritis will go off into the garden and dig holes in which to lie. The adrenergic responses to the failing circulation may trigger primordial fear.

Treatment of phobias

The dog with a phobia has learned to be afraid. The behaviour therapist must help the dog to learn to be less afraid. However, since survival mechanisms like the flight reaction must be indelible if they are to have any value it is probably unrealistic to expect to extinguish a phobia completely, as the following story, my own, illustrates.

In the autumn of 1945 in the yard of Church Farm, Wilden,

Bedfordshire an eight-year-old boy was walking past the bull pen. A herdsman was in the pen giving hay to the bull. Out of the corner of his eye the boy saw the door of the pen swing open and the reddish-brown fur of a Dairy Shorthorn bull rush past the gap at the hinges. The bull erupted into the yard for some 10 or 15 yards, turned, stared at the boy, lowered his head, gouged the deep straw litter of the pen with his right foreleg and charged. The boy fled in terror into the open door of the cow shed past the standings and the feed bin, past the incredulous faces of his step-father and two other workers, out of the end of the byre, across a second yard and behind a heavy wooden door. He remained wedged behind the door until prolonged searching resulted in his discovery. There was much laughter and recounting of the tale. The boy wept and laughed by turns and, armed with a pitchfork, patrolled the yard outside the closed bull pen bravely. In the ensuing years the boy learned to milk cows, lead young bulls by ring and pole. He went to veterinary college. On another farm near Raunds, Northamptonshire, he learned not to be afraid of bulls in the company of the farmer who was ferociously brave.

On one occasion he held two horses whilst the farmer attempted to drive a nasty young Friesian bull out of a field with a tree branch. The farmer suddenly dropped the branch and ran back to the horses, shouting as he sprang into the saddle, 'Come on! You can only bluff 'em for so long. Ride for it!' His own horse skittered about making it difficult to mount. He was barely in the saddle when the bull crashed into them tossing his horse's rump into the air. The horse took a stride or two on his forelegs and then kicked the bull in the forehead with both hind hooves. As they flew down the field to catch up with the other horse and rider the bull stood still, with his nose resting drunkenly on the ground. The feeling was one of elation. There was no fear! The seventeen-year-old thought he had conquered fear.

Eight years later, aged twenty-five and now an undergraduate veterinary student working as a herdsman in the long vacation, he was mending fences on the same farm. Whilst stringing wire in a hedge gap he was spotted by a Friesian bull of notorious temper who was in the next field with some heifers. The bull swaggered across and performed his repertoire of threats, snorting, head tossing and pawing the ground. The young man knew how to deal with this. He took up a fencing post and drove it at the bull with a mighty shout. With unbelievable speed the 1½-ton bull hurtled forward into the gap and wedged there bellowing with rage. The man was paralysed with fear. His heart smashed into his upper ribs as if attempting to exit via his neck. He was breathing slowly with long shuddering gasps. His legs buckled and he almost fell. His sense of time seemed to freeze as he slowly and tremulously turned and crept up on to his tractor to drive away and resume fencing at another spot. He felt sick and shook for some time. In terms of terror he was still eight years old. The phobia was still his. The courage he thought he had acquired was the property of another man.

Apart from emphasizing the durability of phobias (and possibly pointing out the futility of self-defence classes for frightened people) this anecdote tells us something of the way in which we take cues for fear from others. In the presence of a man unafraid of bulls I was unafraid. The persistence of phobias seems to trouble psychological theoreticians. If the taurophobe's family were to shout, 'Dad, there's a bull in the bathroom!' instead of the usual spider which causes him no fear he would probably run away and hide behind the first stout door he could find. A search party of psychologists would want to know why he persisted in running away on the flimsy pretext of having been frightened once forty-seven years before! He would be accused of 'not testing reality' that is not checking in the bathroom to see if there really was a bull. He might be described as exemplifying 'conditioned avoidance' according to the two-process/two-factor model whereby the reduction of fear brought about by escaping or avoiding the feared object negatively reinforces the escape or avoidance behaviour. That is to say he runs away because it makes him feel better. On finding the victim behind the door the posse would squabble about the 'safety signal hypothesis' which maintains that avoidance is motivated *not* by the reduction of anxiety but by the *positive* feeling of safety. To complicate matters, if he were found being comforted by his charming receptionists and veterinary nurses dark murmurings of 'secondary gain' would arise from the older, more Freudian analysts.

In this case the phobia had most definitely been temporarily controlled for the boy by the potent role model of the farmer. Indeed success has been claimed in treating children for dog phobia by the process of 'modelling' in which the children see films and live demonstrations of children handling the animals without fear or harm. Similarly the alarmed dog will seek cues from his companions, canine or human. The canine cues will be the strongest naturally. Dogs can certainly be reassured by the company of other 'steady' dogs, who can also 'catch' fear.

Womble, a noise phobic Labrador, has recently added smoke alarms to her fear list of cap-guns, crow-scarers and quarry blasting, having, it must be said, made great progress with desensitizing to crow-scarers on walks. She recently stayed with Bosun, an unflappable black Labrador and two Border Terriers, Pip and Pickle, for three and a half weeks, her owner being abroad. The weekly smoke alarm tests conducted by the lady of the house and the random tests caused by the man of the house misusing the toaster have sensitized Womble. She was found wedged across a tray of dog food tins in a narrow shelf and the cans had to be removed to get her out. It is now apparent that Pip, the nine-year-old Border male has also become somewhat frightened by the smoke alarms having been in the kitchen with Womble when she was 'denning'. He was never troubled by them previously.

A client of my practice recently reported that his German Shepherd is afraid of thunder and low-flying aircraft. It was not always the case. His neighbour's dog, of the same breed, was thunder phobic and aircraft phobic. He suggested that his neighbour's dog should stay in the house with his to learn not to be afraid. His own dog promptly learned to be terrified.

The first question to ask about such 'contagious' phobic behaviours is – do they have the same indelible imprint as the classic traumatic phobia? Poppy, our paedophobic Jack Russell, was joined after a year by Pod, an amiable male who seemed to acquire Poppy's aversion to children quite rapidly. Neither dog, however, caught their lady owner's astrapophobia (fear of lightning). In the event of an electrical storm all windows were closed, curtains drawn, lights turned off and the two terriers scooped on to her lap on a chair in the centre of the room. The human anxiety did not transmit to the dogs. When the two veterinary surgeons who owned the terriers finally found time to have children, the dogs were eleven and ten years old respectively. Pod's aversion to children rapidly extinguished as the firstborn grew. Poppy's persisted unabated until her death at 15½ years.

It is unfair and wrong to blame all owners for all problems but in the matter of phobias and their reinforcement, rewarding and duplication, the human influence can be massive and absolutely harmful. Typical mistakes may be listed as follows.

1. The apparently irresistible urge to elicit the phobic response. 'Just watch Ollie when I rattle this lid!' When a dog first encounters the vacuum cleaner and shows alarm or makes a playful attack on it few people can resist making a quick pass at the dog with the nozzle and sucking part of the dog's anatomy, fur, tail or even tongue and causing extreme fear. Children are particularly prone to 'tease' in this fashion. If the family dog is alarmed by a radio-controlled toy car once it is likely to be 'pursued' at every opportunity.

2. Failure to allow the puppy (the younger, subordinate pack member) to reaffirm and repeat its greeting and submissive displays as often as it needs. The naturally anxious dog often seems to need to reassure itself in this manner repeatedly. Womble seeks admission to her mistress's workplace every morning with a ritual of greetings. If rebuffed because it is inconvenient she anxiously pads from one forefoot to the other and grins (muzzle wrinkled back, teeth bared) whilst wagging her tail furiously. If the muzzle is held gently she drops on to her back and a perfunctory rub of the stomach settles her for an hour or more.

3. Showing extreme anxiety or anger towards the dog which is itself in a state of anxiety. The response of Gemma's mistress is classic. Gemma's general anxiety symptoms, which did not respond to valium, phenobarbitone or amitriptyline, rapidly eased and almost ceased once Mrs S. became calm and stopped reacting angrily and fearfully herself. The dog which is anxious or chronically afraid is alert and aroused for the purposes of flight on cue. It may respond to any cue, not merely the established phobia; men with beards, umbrellas, the behaviourist's clipboard, blue trousers, white coats are familiar examples. All handling and interactions with the anxious dog must be conducted extremely carefully in order not to trigger fresh fears.

Factors affecting treatment

Any treatment plan must take account of all the factors discussed so far in this chapter. The breed of dog must be considered. Different parts of the brain and subtly different combinations of chemicals in them deal with differing types of fight and flight behaviour. Prey objects release predatory aggression, predators release fear aggression, members of the same species release internal aggression, threats to the young release maternal aggression and intruders release territorial aggression. Selection by breeders has emphasized or exaggerated the strength of these brain programmes for special purposes.

The socialization within the litter and the new home need to be taken into consideration.

Old fears must not be renewed and new fears must not be acquired. The therapist needs to know whether the subject is a normal dog with a focal phobia or an anxious dog with a repertoire of phobias and the ability to develop new ones unexpectedly. The natural psychology of the dog must be accommodated, e.g. a den or safe burrow may be needed.

Before phobia therapy is undertaken the 'level playing field' of sound health must be confirmed. A full neurological examination should be made and liver and kidney functions checked.

In order for behaviour therapy to proceed it may be necessary to attempt to control or at least reduce the worst effects of fear by medical means. Fear and its pharmacology must be understood.

The physiology of phobia

In an emergency the dog's body is primed in an instant for either attack or defence, 'fight or flight'. Extra amounts of hormones (adrenaline and noradrenaline, also known as epinephrine and norepinephrine) are excreted from the adrenal medulla. The sympathetic nervous system, which is 'wired' to the entire body, signals the emergency to every organ. The resting blood supply to the body parts is suddenly diverted away from non-essential areas such as the skin and intestines to the muscles of the skeleton by constriction of blood vessels in the non-vital area and dilation in the useful area. The blood is pressed into a smaller volume and the heart begins to pump more quickly and with larger beats to raise the rate of flow of oxygen and sugars to the muscles. The oxygen supply is ensured by the widening of the bronchioles of the lung and more rapid breathing. Red blood cells are rushed into the bloodstream from the reserves in the spleen. The blood sugar rises. Sensitivity to pain falls and the metabolic rate rises. The eyes 'widen' as the pupils dilate and the upper eyelid muscles retract. The dog can now either attack or run for its life with tremendously increased efficiency.

If the fear responses are excessive, however, things can go wrong, as Gemma showed. She had the classic 'red eyes' due to her raised blood pressure congesting the scleral blood vessels on the white of her eye and they seemed to bulge because her upper eyelids were drawn tightly back. Instead of being 'dry mouthed' with fear she was drenched with saliva. In some dogs in Gemma's state, the bowel and bladder sphincters fail, while at a lower level of fear stimulation they close tightly. These events are the measure of extreme fear in which the emotional centres in the brain stimulate the sympathetic nerve centres so violently that the effect spills across to the parasympathetic centres, which send contradictory messages to the salivary glands and anal or bladder sphincters. If the breathing becomes too deep (hyperventilation) human patients report dizziness, fainting and tinnitus.

Dogs certainly topple over and seem to faint. But might they also have tinnitus? This could explain the curious barking at invisible objects they can hear but not see. Finally humans in severe phobic attacks can have partial-complex seizures which resemble those in dogs.

The pharmacology of fear

Veterinary neurologists group behaviour disorders into four classes according to the preferred therapy.

Class 1. Conditions for which progestagens are the first choice, such as aggression, urine-marking, mounting, sexual perversion, roaming and spraying.

Class 2. Stress-related conditions such as phobias and anorexia for which benzodiazepines are the first choice.

Class 3. Anxiety-related conditions such as excess grooming and separation anxiety in which amitriptyline is now the first choice and progestagens used if this fails.

Top: Puppies need to get used to busy street noises

Above: It is important that they socialize with children

Left: An early visit to the pub will help in the socialization of the young puppy

Right: A destruction-proof area and also a comfortable den

Left: Digging is this dog's way of coping with anxiety

Above: Destroying the cushion may seem like great fun, but it is really indicative of stress

below: Desensitizing a car phobic Springer – in the car with the window open, engine not running

right: Don't over-comfort the thunder phobic dog during storms, to avoid learned dependence

Early relaxation during controlled exposure to hot air balloons for a phobic Bullmastiff *left*. He also has the comfort of another, confident, dog (*above*)

Right: Playpens in busy places are invaluable for early socialization of puppies in rescue

Below: Dogs can be taught to accept being groomed while in kennels, but treatment for behavioural problems has to be done when the dog is in a home environment

Above: Making the dog too comfortable might cause problems later

Left: Separation problems are frequent among rehomed dogs. Timely advice can prevent or cure the problem, breaking the cycle of dogs being passed from home to home

Below: A ten-week-old puppy learns the hand signal to sit

Above: Teaching the recall is easy with the right incentive

Below: Learning the down without pressure

Left: Plenty of handling and gentling of kittens between two and seven to eight weeks is essential if they are to make good pets

Above: Wool-eating Siamese cat

Right: Cats prefer soft rakeable substrates to use for toileting – soil is usually ideal

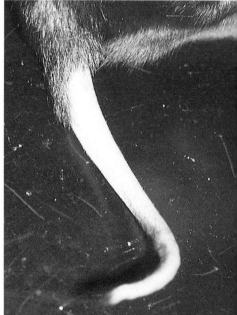

Below: Self-mutilation – fur has been pulled from the tail by a cat traumatized by the arrival of a new baby in the household

Above: The human–cat bond can be very strong
Right: Frisky – RIP

FRISKY

Class 4. Hyperkinetic syndromes in which dexamphetamine is specifically indicated (this being a controlled drug).

Intervention with drugs can be dramatically successful in some cases. This is enormously valuable where the problem is so severe that the owner is distressed beyond the point where long-term slow improvements have much appeal. The danger is that the busy veterinary surgeon and the desperate owner may take the easy way out and simply use the drug instead of addressing the behaviour problem.

Acepromazine Maleate (acetylpromazine or ACP)

ACP is the most widely used of the phenothiazine ataractics (greek *ataraxia* – not disturbed). Veterinary practitioners have some thirty years' experience of its use as a sedative, anaesthetic premedicant and anti-emetic. There is, however, a universal tendency to use too high a dose for the effect required and to time the dose too late for the effect to be achieved.

An oral dose of 1–3 mg per kg produces deep sedation in the dog. This is accompanied by sleepiness and reduced motor activity leading to wobbling hind legs in particular. The facial expression is dramatically changed. The skin of the forehead becomes loose and wrinkled, the upper lids droop and the nictitating membrane (third eyelid or 'haw') is relaxed and protruded partially across the eye. The dog may lie down and become somnolent. The effects usually wear off within three to four hours but may last for seven hours. The effects of ACP often seem to be much more dramatic in the more nervous dog. If the owner of the dog is already anxious about a severe anxiety problem in the animal, the effects of heavy sedation are in themselves so alarming that the treatment is abandoned with the highly predictable comment, 'I'll never give him that again!'

It is of the utmost importance that the dose of ACP intended as a buffer or block against panic is given *before* the animal is frightened. It is not possible to 'overtake' severe phobia once it is elicited. Attempts to treat thunder phobia by repeated incremental doses after the fear has begun result in an alarmingly sedated dog (and a thoroughly frightened client) the following morning. The preventative dose for phobia must be given well before the stimulus or trigger is encountered. It must also precede any cues that may alert the dog. Preparations to leave for the veterinary clinic or a car journey must be made after the dosage, not before. Thunderstorms and the 'electric' atmosphere that precedes them are difficult to predict but listening to the weather forecasts can help. When Gemma finally demonstrated her thunder phobia ACP was immediately supplied to be used as a defence.

Her owners have managed to cope by giving a small dose daily in advance of a forecast storm and a dose immediately a storm seems imminent. Gemma's fear has been greatly reduced to the point where, although she is clearly not 'happy', she will go and lie quietly in a den provided for her. For 5 November fireworks, which seem to be exploded for two weeks before and a week after the traditional date, a regime of small doses at six-hour intervals can be highly successful during the vulnerable period.

One type of car phobia seems to be associated with acute nausea. 'Bomb proof' Bosun has always been a most enthusiastic car passenger and untroubled by the motion. His litter mate Ben was collected from the breeder's house on the same day in the same car. They completed a journey together and arrived in a very different condition. Bosun sat on my lap and was unperturbed. Ben lay on the back seat as his new owner was driving and was very sick. Almost immediately after vomiting he voided his bladder and bowels and showed great distress. For two years afterwards, whenever he went in the car the same thing happened. Latterly he is rarely sick (he takes 'Sea Legs') but lies very quietly on the floor and makes no attempt to look out of the windows.

If the highly developed vomiting ability of some species is of survival value it seems reasonable to expect it to rank alongside pain and fear as a 'memorable' stimulus, i.e. with useful aversive properties. A recent case of rat bait ingestion was admitted to the clinic, given apomorphine and made horribly sick. The next day the dog was most unwilling to re-enter for antidote therapy. The aversion diminished after a series of uneventful handlings and injections. The inference might be permissible that inexplicable aversions or phobias where nothing has been observed can be associated with pains, frights or *episodes of nausea* which were not noticed or ignored at the time. Even if we accept no more than a coincidence of nausea with fear in some dogs then ACP is certainly the drug of choice (given well in advance) for car travel.

Diazepam (valium)

This drug is increasingly used in canine behaviour therapy. In the treatment of human anxiety it has been found to reduce apprehension without decreasing alertness or understanding and learning ability. For this reason it finds theoretical favour among animal behaviour therapists. There is however the inconvenience of considerable variation in response. Some dogs become excited while others are not sedated. Some authorities consider there to be no tranquillizing effect in healthy dogs. For reliable tactical control of severe phobic responses ACP would seem preferable.

Apparent *disinhibition* or the release of suppressed behaviour can be highly inconvenient or even dangerous. It may be that dogs that have been *controlled* or trained by punishment have suppressed behaviours that they are afraid to indulge. The anxiolytic may remove the fear that controls the dog. Behaviourists should be alert to problems arising with sedation in dogs whose inhibition from undesirable behaviour has been established by the *startle and command* method. Here the sudden adrenergic response to training discs or other source of startling sound has aided learning. Unexpected aggression has been reported with all of the minor and major sedatives. Reversion to 'puppyhood' habits of wetting and slipper chewing etc. have been reported.

Beta-adrenergic blocking agents: a possible application

The speaker at the Annual General Meeting of the Worcester branch of the RSPCA attempted to deliver a talk about pet ownership. He normally experienced a mild increase in heart rate when about to speak before an audience. Suddenly his heart began to pound massively and very fast. His mouth went dry and his voice became a hoarse whisper. His legs would not support his weight. He had to talk sitting down and in a sort of croak. After about twenty minutes he finished his talk and slumped in his chair exhausted. After a further fifteen minutes he was able to stand and walk to the tea room wearing an extra borrowed coat because he felt freezing cold and was shivering. The speaker had experienced a severe adrenergic attack. The symptoms were all the somatic or bodily events associated with extreme fear. Two days before the man, who was a coronary heart disease patient, had been taken off a daily dose of the beta-blocker Atenolol. (Sudden cessation of such treatment is not recommended. The doses should be reduced gradually.)

During three months on Atenolol he had noticed a remarkable calmness and lack of worry. After the adrenergic attack he was astonished to discover that an array of nervous tics, face and nose twitching, head jerking, blinking and funny little coughing noises in the throat (which had troubled him as a child and had remained in a much muted form through adult life) had returned with a vengeance. The patient now realized that he had been prone to anxiety since eight years of age when he suffered rushing, whirling, vertiginous sensations in the night with numbness of the hands and face. Whilst his mother and step-father screamed and fought downstairs he was cringing in his bed hyperventilating! A resting level of anxiety about anything and everything from anticipating double yellow lines outside the house to practice profits had combined with minor phobic responses (tachycardia, coughing) to certain events, family rows, angry telephone calls, etc. The beta-blocker had wiped all this out.

The calming effect of beta-blockers has been known for twenty years but their employment in anxiety therapy has been limited to controlling the somatic effects (tachycardia or tremor) in some anxious patients. There is, however, widespread use of tactical dosage prior to giving lectures, encountering hostile audiences and playing snooker for money!

Elsa: a clinical note

After two attendances at the clinic for ear treatment during which Elsa, a two-year-old Golden Retriever bitch, panted, struggled, jumped and swooped about the room trying to escape, a trial with Atenolol was suggested.

Elsa is a generally 'nervous' dog who shows apprehension of people in the street (growling at hat wearers) and positively panics at the veterinary clinic.

A morning dose of 25 mg Atenolol was given in tablet form for one week. Elsa weighed 36 kg.

During the week Elsa slept 'a lot more', was not 'so boisterous as usual' and instead of bolting her food 'ate some and then came back and ate some later'. Out in the street Elsa still growled at people wearing hats but her behaviour at the clinic was much better. She still looked worried but gave only a token struggle when her ears were examined. Her owner was most impressed: 'Normally I'm exhausted by the time I leave here!'

Much more evidence is obviously required but if Atenolol can restrain fear, preventing it from escalating into blind panic without blunting perception, a means may be available to treat dogs in circumstances where the phobic stimuli cannot be avoided.

Phenobarbitone

Occasionally people describe episodes of disturbance in the night. A dog will start barking and 'worrying' for no apparent reason. After a series of such nights and no discernible pattern or explanation a course of phenobarbitone sedation for seven to ten nights often seems to solve the problem.

Valid objections to attempting to teach drugged dogs to manage their anxiety or control their fears are that learning ability may be impaired and the sensation of being 'drugged' becomes an essential part of any responses that are learned. The matter is clearly complex, but it seems reasonable to use drugs to prevent the dog from being utterly overwhelmed by fear. On the other hand, it might seem a bad idea to block the effects of adrenaline too thoroughly given that a little fear and the closely allied emotions of excitement and enthusiasm can contribute so much to learning.

Learning to be less afraid: systematic desensitization or counter-conditioning

In essence this treatment method aims to convert the dog's fears, step by step, into pleasure using the powerful urge of hunger or the drive to play or the pleasure of sheer contentment to replace the fearful response. The very best type of treatment programme incorporates as many pleasurable substitute responses as possible. Enormous care is taken not to undo progress by evoking fears. Great patience is needed.

The dog is studied very carefully for a few days to find out what elements of the daily routine elicit most pleasure or enthusiasm. Cues that set the tail wagging will be those that signal imminent feeding, invitations to play with favourite toys or the arrival of favourite people. Thought may be given to heightening or sharpening 'appetites' by some very subtle rationing or 'deferred' gratification in the matter of access to toys or over-generous meals.

In a programme designed to desensitize a dog to noise, for example, a range of recorded sounds of the type of which the animal is phobic is presented by steps and in increasing volumes.

The session begins with the dog lying (by choice) and relaxed without fuss. An unseen hand turns up the volume of a stereo disc or tape player *slowly* until the dog registers the sound by twitching or raising *one* ear (both ears and head up is too fast). The dog must recognize the sound but not be startled.

Immediate difficulties arise with dogs that discriminate between noises generated electronically within the house and external (real) noise. Womble appeared at first to be 'tape proof' possibly due to having 'sat through' countless video and television films of modern histrionic mayhem at high volume. Some patience was required in taping a selection of 'bangs' and varying tone and volume until the ears eventually flapped at a level likely to alarm the neighbours. Substitute thunderstorms may present problems if the stimulus does not have the quality of the real thing. One authority suggests 'excellent video equipment and recordings, darkened rooms, strobe lights to simulate lightning and lots of 'bass' tone turned up' until the dog registers fear! This offers a battery of potential mishaps and is not recommended here.

Having gained a response in a susceptible dog the exact volume setting is noted or marked and the sound turned off. For the next three days the tape should be activated just before the dog is awarded one of the pleasurable experiences discussed. During the excited or enthusiastic response the tape should continue to run until the 'fun' begins to subside.

This routine can enter a second phase in which the volume is increased every day. Should the dog show the slightest 'suspicion' the level is carefully noted and maintained for three days without increase. The goal is a dog which starts wagging his tail enthusiastically when he hears the noises.

A major feature of this method is the third stage which moves the whole process out of the house in order to help those dogs which are afraid to go out at all or whose walks are spoiled by noise phobia. In essence *indoors* must become quite dull and unexciting in order to contrast with the fun that is to take place. Attention, fussing and playing in the home must dramatically reduce or cease. The dog and owner venture forth with a suitable tape player, a favourite toy and a complete meal in a bag. Once outside the tape is played to elicit anticipation of something pleasant. In this frame of mind the dog must walk some distance before being fed, fussed and enjoying a brief playtime with the toy. Back home the dog must be ignored for a time to reinforce the contrast. The daily sessions extend the time spent walking and listening to the tape before the rewards are given. By now all of the dog's daily food is eaten on the walk.

The fourth stage is enacted in novel situations or areas unknown to the dog and therefore not associated with fear. The tape is not played continuously as the noise is to be used as an isolated *incident*. At some point on the walk the tape is played and the rewards dispensed. Titbits can be employed and normal meals reverted to the home. Several such *incidents* are repeated on each walk.

Stage five consists of cautious returns to familiar areas where associations with fear have been made. Exposure to the tape and rewards are repeated. *Apart from remembering previous fear inspiring events the dog will have a mental map of the area and may seek to hurry the walk to its termination or 'cut back' to the car.*

If the dog is startled to any degree the companion human must not react or 'cue' the dog. The tape should be played at once and when the dog relaxes a reward given. It is vital not to reward the apprehension itself or allow fear to be the cue to terminate the walk.

Summary

Much of the behaviour counsellor's work will be with manifestations of the 'fight or flight' mechanism either in its extreme forms of aggression and phobia or its mid-stations of excitability and anxiety. Many of the symptoms of distress are activities which relieve the stress or tension. Tics, obsessive movements, nail biting, ear twiddling, arm scratching, paw chewing are all means of gaining real relief from distress. They belong, along with hysterical laughter and paroxysmal weeping, among the mechanisms by which the mind relieves itself from tension. (Fifty years ago nail biting was recognized as a 'motor discharge of inner tensions' due to emotional difficulties. A quarter of nail biters in one survey showed motor restlessness and sleep disturbances, one-fifth suffered from tics. The affliction was seen to spread among whole classes of schoolchildren *disciplined by a particularly severe teacher!*

There is a tendency in the medical field to *label* each type of relief behaviour and address it as a disease: obsessive compulsive disorder, acral lick dermatitis, stereotypic dyskinesis and so forth. Drug therapy is then applied to the *disease*: clomipramine for paw chewing or amitriptyline for separation anxiety. The behavioural investigator must discover the source of the symptoms and the human influences making it worse in some cases.

Our taurophobe discovered as a child that grimacing and blinking, and above all fiddling with his ears, relieved tension. As an adult employing relaxation techniques for general stress control, having *insight* which has enabled him to identify and extinguish some minor responses and having side-stepped the *bulls* by specializing in pets he was delighted to find twenty people at his T'ai chi class happily stroking and squeezing their ears (his wife used to scold him, but now it is *legitimate!*). It slows the pulse rate in dogs as well as humans.

Partial success is often the outcome and can be satisfactory. Womble's owner has not tried fully to desensitize her to bangs. Walks in different areas in the company of a dog or two humans who do not react to noise themselves, and rewards for returning when whistled or called, are having a very good effect. Candidates for work with the gun who are frightened of the noise must by and large have only the focal phobia if they are to be desensitized. A similar problem exists with police dog recruits. Occasional success is achieved. Tex was gun-shy but in every other respect an ideal dog for police work. A graded programme of exposure was begun by arming the handler's sons with cap-pistols. The dog successfully developed tolerance from toy guns to revolvers and finally to powerful shotguns. He was not an anxious type and his enthusiasm for his handler, the family and the job supplied the unstructured counter-conditioning or reward. Few cases will be as easy.

A firmly structured life with calm, consistent behaviour by the adult humans in a family can provide dogs (and children) with the opportunity to relax and feel safe. From this haven the individual ventures forth to exercise or rehearse the entire range of appetites, desires and activities for which it is programmed. If the environment does not provide the real thing a substitute is to be found in play. Mental and physical health are indivisible and dependent upon such rehearsal. Lack of opportunity to exercise the brain's programme can lead to misery. It holds true whether you are a dog or a philosopher.

Suggested further reading

T. A. Betts and A. Blake, 'The psychotropic effects of Atenolol in normal subjects: preliminary findings', *Postgraduate Medical Journal*, Vol. 53 (suppl. 3), 1977, pp. 151–6.

N. H. Booth and L.E. McDonald (eds.), *Veterinary Pharmacology and Therapeutics*, Iowa State University/Ames, 1988.

R. D. Gross, *Psychology: The Science of Mind and Behaviour*, Hodder & Stoughton, London, 1987.

P. Neville, *Do Dogs Need Shrinks?* Sidgwick & Jackson, London, 1991.

V. O'Farrell, *Manual of Canine Behaviour*, BSAVA, 1992.

J. Rogerson, *Understanding Your Dog*, Popular Dogs, London, 1991.

E. Slater and M. Roth, *Clinical Psychiatry*, Ballière Tindall, London, 1977.

M. J. Swenson, (ed.), *Duke's Physiology of Domestic Animals*, 8th edition, Cornell, London, 1970.

V.L. Voith, 'Behaviour problems', in R. Chandler (ed.), *Canine Medicine and Therapeutics*, Blackwell, Oxford, 1984.

J. Z. Young, *Programs of the Brain*, Oxford University Press, 1978.

The author would like to give special thanks to Peter Neville for his help during the writing of this chapter.

8 Behaviour Problems in Rehomed Dogs

Gwen Bailey

Why are dogs given up for rehoming?

Blue Cross, one of the largest animal welfare charities in Britain rehoming unwanted dogs, conduct an ongoing survey to find out why people give up their pets. Each year, all the reasons why people bring their dogs in to branches of the Blue Cross are collected and analyzed.

It has been found that one-third of adult dogs are given up by their owners because of an uncontrollable behaviour problem. Moreover the frequency with which dogs are given up for reasons such as 'owner moving', 'owner can't cope' and 'stray' (a total of 26 per cent of adult dogs) leads us to believe that well over half may have been given up because their behaviour was less than perfect. The survey also shows that dogs exhibiting behaviour problems are much more likely to have had many different homes than well-behaved dogs. It is not uncommon for noisy or destructive dogs, or those that do not get on with other pets, to have passed through as many as six homes.

In an effort to address some of these problems, Blue Cross became the first charity to employ a full-time animal behaviourist to help its shelters rehome more dogs more successfully. We have made a start initially at our branch in Burford, Oxfordshire, where we have found that the number of dogs returned to the shelter has been dramatically reduced. If this holds true generally, it means that the number of successfully rehomed dogs can be greatly improved. The success rate for this branch during the first year of its operation, where all of the following ideas were put into practice, was 93 per cent as compared with the norm of only 50–80 per cent at many other rescue kennels.

Preventing 'problem' dogs from being given up

Dogs which come in to rescue kennels with an existing behavioural problem are much more difficult to rehome and tend to stay longer and use up more of the shelter's resources. Once homed they are also

93

much more likely to be returned. However, if the owners were counselled and provided with expert advice it is likely that the dog would not have to be brought in at all. The dog would therefore not undergo the very stressful experience of being rehomed, the owners would be educated, which may well prevent another puppy being brought up with similar problems, and valuable resources conserved that would leave us free to help more animals.

The original owners of the dog are the best possible people to cure any problem behaviour, not only because they have all the answers to why the dog has that particular behaviour pattern (which is the key to the cure), but also because they are likely to have more motivation for correcting the problem. Many owners that bring their dogs in for rehoming are not irresponsible or uncaring, but simply at the end of their tether and lacking the knowledge and understanding to put things right. More often than not, they are more than happy to work through a cure once they are given the knowledge of how to do so.

One such case was that of Pepe, a five-year-old Dachshund. His 86-year-old owner had contacted Blue Cross to find out about the possibility of rehoming him since she could no longer cope with his behaviour and did not want to have him put to sleep.

Pepe had established himself as head of the household when his owner's sister, who had lived with them at the time, sadly died. Pepe noticed the change in circumstances, and while his owner was still vulnerable seized the opportunity to take control.

This, in itself, was not too much of a problem, except in regard to the use of the phone. His owner was becoming increasingly frail and more and more dependent on the use of the telephone to summon outside assistance. Normally, Pepe was the centre of his owner's attention all day every day – apart from when she was talking on the telephone. Pepe took great exception to this intrusion into his world and refused to tolerate it. He began to attack the legs of the table on which the telephone stood, being too small to get at the telephone itself. When this failed (his owner stood the table legs in tin cans so that he could not damage them!), he tried to bite and scratch at the carpet. This was ignored so he tried a loud, high-pitched yell whenever his owner tried to speak. This resulted in the receiver being quickly replaced. Pepe also began to guard the telephone and would do so for up to two hours after a call.

Following a visit to the home and a consultation with the owner, the cause of the problem became clear. By removing many of the privileges he had previously been allowed, usually reserved only for high-ranking dogs, such as being fed first and being allowed to take food off his owner's plate, and by gradually building up periods with no attention being given to him by the owner, we were able to demote him sufficiently for him to rethink his position in the pack. His owner was unwilling and unable to do all the things necessary for a complete cure, but we were able to take enough edge off his superiority to reduce his telephone aggression. We were then able to use aversion methods to prevent him barking during calls.

Eventually we reached a point where the owner no longer needed to give up the dog. He was his owner's great friend and protector, and it would have been very distressing for her to give him up. It would also have been very difficult to find him a new home. At five years old, and having had all his own way all his life, we would have needed to find a very dedicated new owner who was willing to change him, or a very tolerant one. The stress the dog would have had to go through in order to adjust to a new way of life would also have been considerable. By keeping the dog in the original home, all of these problems were avoided.

There will always be owners who simply do not care enough to sort out their dog's behaviour problems; they just want to get rid of the dog and start again. If owners are not interested in trying it soon becomes apparent. In these cases the most that can be done is to gather enough information about the dog to make a diagnosis so that the correct treatment can be ready for the new owner to work through as soon as the dog goes into its new home. In such cases, we have found that it is important to see the dog in its original home surroundings. Some owners are reluctant to give details that they consider will jeopardize their dog's chances and conceal important facts. Seeing the dog on its home ground can be very revealing.

There are also some cases where the dog–human relationship has broken down completely and no amount of counselling can help the situation, or where the home environment is such that the prescribed treatment stands little chance of working. Fortunately, these are in the minority. Collecting sufficient information, finding a suitable new owner and giving them appropriate advice so that the problem behaviour does not reoccur in the new home is the answer in these cases.

Assessing dogs as they come into the shelter

All those who work in animal shelters know that not all owners are entirely truthful about their dog's behaviour when they bring it in to be rehomed. They often consider that it stands a better chance of finding a new home if they conceal certain facts, such as that it is destructive when left alone, or bites postmen or chases children! Because of this, we are beginning to develop ways to get to the truth in cases where we suspect we are not getting the whole story.

Owners bringing in dogs for rehoming are asked to complete our standard admission form. This asks a number of general questions about the character of the dog, one of the most important of which is 'Can you groom the dog?' If there is any uncertainty about the dog's temperament, owners are asked to demonstrate so that the staff can assess how well this can be done. Many of the 'problem' dogs seen by behaviour counsellors are ones that see themselves as higher ranking than the members of the family with whom they live and who have become aggressive towards them as a result. Usually such dogs will not tolerate being groomed and so this gives us a good indication of how dominant the dog is with its owners. We can then assess the dog, pair it with suitable new owners and give appropriate advice when the time comes.

If a behavioural problem is suspected further details are taken by the behaviourist once the admission form is completed. This gives us a more detailed knowledge of the dog's past experiences and allows us to give the special care and advice the new owners need to ensure that the problem does not surface again.

We are developing a new admission questionnaire to be completed by the original owner at the time of admission. This leads them through a series of questions designed to allow us to spot potential problems that may otherwise reoccur in the next home.

Problem dogs in kennels

Unfortunately, there is nothing that can be done to correct behaviour problems once the dog comes into kennels. It is impractical in terms of time and facilities, and it is also very necessary for the dog to have a real owner that it can relate to while the problem is being corrected. There is also no guarantee that a behaviour problem corrected in one environment will not reappear in another.

Nevertheless, a few small things can be done to make things easier for the new owners, such as teaching nervous dogs to enjoy being handled, or winning games with dominant dogs and getting them to accept handling on human terms, but, in general, all behaviour modification needs to be done in the new home.

At the moment, most dogs are chosen by their new owner on the basis of appearance and appeal. This seems to be successful in the majority of cases, but occasionally people will choose a totally unsuitable dog for their characters and lifestyle and have to be dissuaded from taking it. At the moment, we are looking into ways of assessing dogs so that we can provide a character profile for each dog in the shelter. The potential owner would then be encouraged to read the profiles of all the dogs in the shelter before going out to look at them. Then, as they look around the kennels, they will see the dogs that on paper are more suitable for them and, hopefully, be more likely to choose one of them.

Educating new owners

Before new owners are allowed to take their chosen dogs into their homes, they have to attend one of our lectures on dog behaviour. This helps them to understand their dog better and to set the ground rules for their new dog. Common problem behaviours are described and explained and preventative measures outlined. This is particularly useful for dominance/hierarchy and destructive problems. Owners are then told that full support is available and advised to contact us at once should a problem develop, rather than leave it until a problem behaviour is well established.

The disadvantages, as well as advantages, of taking on a rehomed dog are spelt out so that new owners are not under the misapprehension that they are giving a home to a 'perfect' dog. Problems are not over-emphasized but neither are they glossed over, and we try to give a balanced, realistic view of what life with the new dog will be like.

Very occasionally a potential new owner decides against taking a dog after the talk. However, this is rare and we feel that it is better that they make up their minds at this early stage rather than deciding it was not such a good idea after they have acquired the dog.

The main points covered in the lecture are as follows:

1. Why the new owners have been asked to attend the lecture, i.e. that dogs are a different species from us and that by understanding some of the differences, and by learning some guidelines, they can start to build the correct relationship with their dog as soon as they get it home.

2. How dogs are related to wolves and that their behaviour is very similar. This is used to explain the concept of hierarchy and the importance of the human being the dominant animal in the pack. How the new dog will have a predetermined view of humans depending on the genetic make-up and his previous experiences, but that how he is treated in the first two weeks in his new home will make all the difference to how he sees his role in the new family.

3. How to ensure that the dog learns that he is bottom of the pack is carefully explained with special reference to the following: sleeping areas; order of feeding; grooming; and control of games. These are explored in detail and a lot of emphasis is placed on this aspect of dog ownership. How to ensure that the dog sees itself as being below any children in the family is also carefully explained.

4. The importance of ignoring a new dog unless you have chosen to interact with it is explained. Owners are told that not only is it a way of acquiring high status, but also it helps those dogs that tend to become very attached and dependent on their new owner.

5. Teaching the new dog the house rules. Reward (how to use it and when) and punishment (how not to do it, and why it often doesn't work) are looked at. Why delayed punishment does not work and the use of environmental correction are also covered.

6. Chewing (one of the most common reasons why dogs are returned to us). The main reasons for this are explained and advice given on how to prevent it happening.

7. Dogs and children/babies. Advice on introductions is given and on how to overcome some of the difficulties that may be encountered.

8. House-training. How to teach dogs that their toilet is in the garden is explained. Importance is placed on allowing the new dog time to adjust its body to its new routine, which will be very different from that in kennels.

9. Some advice is given on how to deal with any problems that may occur during the first few nights. Some dogs going into a new environment are likely to react badly to being left alone during the first few nights and owners need to know what to do to ensure that barking, whining or scratching at the door do not become bad habits.

10. The importance of adequate exercise is stressed and how the needs of individual dogs will vary.

11. Owners are told to expect any problems that their new dog may exhibit to arise approximately two weeks after they acquire them. Dogs tend to behave well for the initial two weeks and lull their owners into a false sense of security. If the owner is watching out for any signs of trouble towards the end of the second week, they are less likely to be surprised by any occurrence of problem behaviour.

It is stated several times throughout the talk, and at the end, that we are there to help with any difficulty, no matter how small and that it is far better to sort out the problem straightaway than to let it become an unacceptable habit.

The talk is supported by a visit to the home where the dog will be living and any questions that have arisen as a result of the talk are answered.

Advice for minor problems will be given at the time of the home visit. Potential problems will have either been noted from the original owner or from the behaviour of the dog while it has been in kennels. This information will be passed on to the new owner and advice given on how to prevent these problems from arising in the new home.

It has been found that it is easier to sort out behaviour problems which have arisen after the dog has been in its new home for a while with people who have taken the advice given at this talk. This is because the owners have built the correct relationship with the dog from the start. The talk is a very low cost part of dog behaviour work and requires just one hour per week. The benefits seen make it very definitely worthwhile.

In one case where a new owner kept to the advice given in the talk quite a serious problem was easily overcome with a fine adjustment of the dog's environment. The new owner in question took on two German Shepherd Dogs that had come into our shelter as a pair. As so often happens when two dogs are kept together, the younger of the two had formed a stronger bond with the older dog than it had with the human members of its pack. This, added to the breed's natural wariness of people, resulted in quite a deep-seated mistrust of people. After the dogs had been in their new home for a few weeks, there was an incident where the owner opened the front door in order to take the dogs out for their walk at the same time as someone was running through their open-plan front garden close to the house. The younger dog bit the runner and the owner sought our advice on how to prevent such an event from occurring again.

Due to the fact that the owner had taken on board all our recommendations, he had already built a very good relationship with both dogs. Both dogs considered him to be very high-ranking and so he had good control over them. He had done this by using our dominance guidelines rather than force, and so both dogs trusted him completely. Also he had played a lot with each dog on its own so the younger of the two had begun to form a strong bond with him thus diminishing his mistrust of humans in general.

Since the process had already been started and the 'problem' dog already had the correct relationship with his owner, it was a relatively simple matter to arrange for food and games to come from other people so that the dog began to see them as a source of rewards rather than a threat. As his attitude to people changed he became gradually more trustworthy and he now no longer behaves in a territorial manner.

Had the owner not had any advice, it is likely that he would have used force to control these two lively dogs, deepening the mistrust of the younger. If he had not played separately with each dog the bond between dog and owner may have been less strong. So much would have had to have been done that the owner may well have found the task too daunting and returned the younger dog. As it was, both dogs settled into their new home and have become good family pets.

Special care for known problem dogs

When a dog has come into the Blue Cross with known behavioural problems, the home visit is done by the dog behaviourist to help the new owner fully to understand how the problem behaviour came about and what to do to prevent it from occurring in their home. This approach has been particularly useful in curing many behavioural problems, especially those relating to dominance.

One such case is that of Fred, a three-year-old Bulldog. Fred had taken control of his elderly owners, who were covered in scars from Fred's bites. Not only did they find it very difficult to change the way they had always treated him but they had also become frightened of him. Finally, they reluctantly decided to give him up as there was no likelihood of improvement in the situation.

In many shelters, Fred would simply have been labelled as 'dangerous' and put to sleep. However, it was clear from the information we collected that Fred was just a normal dog who had been led to believe he was leader of the pack. He bit his owners when, in his eyes, they tried to act like his superiors. He was simply trying to keep control of his pack.

After an assessment period in our kennels, during which time we tested his reactions to the staff's attempts to dominate him, we decided to try to place him in a new home. We looked for owners with strong characters who had had previous experience with similar breeds.

Luckily we found such a couple quite quickly. Before they took him, they were given a complete set of house rules to follow so that Fred would know, as soon as he got home, that he was the weakest of the pack members and his place was at the bottom of the pack hierarchy. His new owners followed our instructions to the letter and Fred settled down without any problems. There were one or two challenges after the first two weeks, but the owners had been instructed on how to recognize them and easily dealt with them.

After a few months, the owners were so pleased with his progress that they rehomed another dog from us as company for him.

Obviously, not all cases have such a successful outcome as this one, but the fact that the majority do encourages us and makes us realize that without such a service many dogs would be put to sleep unnecessarily, or attempts to rehome such dogs would end in failure, which is not good for either the dogs or the families that take them. By tackling the problems, many such dogs can go on to lead useful, happy lives as family pets.

Finding new owners for problem dogs: is it ethical?

Dangerous dogs or those where there is little chance of recovery are not passed on.

Owners are always told if a dog has a particular problem as soon as they show an interest in it, before they become emotionally involved. If they do decide to take it on every effort is made to give them help, advice and support as they work through the problem. They are always told they can return the dog to us should the venture fail.

Not every case is successful, of course, but careful matching of owners to the particular character of the problem dog increases greatly the chances that a case will have a fruitful outcome, and is definitely worth trying for the dog's sake.

This leaves us with the question of what we would do with dogs with known problem behaviours if we didn't follow the lines we do. The quick and easy answer is, of course, for them to be destroyed, but few charities, us included, would survive if their supporters knew that a large proportion of their intake were put to sleep. The shelter could simply refuse to take problem dogs in, but they would get some anyway because people do not always tell the truth, and then what would we do with them? We have decided that the only responsible way is to face up to the problems, and do something about them.

Whenever possible, the original owner is encouraged to take responsibility for their dog's behaviour and do something about it. When this is not possible and owners abandon their responsibilities, the shelter, and ultimately the person offering the dog a new home, takes on the problem.

Preventing the return of rehomed dogs

No matter how hard we try, some dogs develop a behavioural problem in their new home, but by providing help, advice and support to their owners, it is possible to keep dogs rehomed. Often just a little advice over the telephone is all that is needed, although sometimes an in-depth discussion of the problem is needed together with a programme of treatment.

Separation anxiety problems, such as chewing, barking and house soiling are particularly common in rehomed dogs. They are, unfortunately, just the sort of problems of which new owners are particularly intolerant. A little advice as soon as the problem appears can both alleviate the trouble and reassure the owners that they have not chosen the 'wrong' dog. Separation problems may well be exacerbated or even caused by rehoming and it is therefore particularly important that dogs exhibiting these behaviours are kept in their new homes long enough for them to settle down and stop exhibiting them.

An example of such a case is Purdy, a three-year-old Pointer-cross. Purdy came in for rehoming because of a marriage break-up. She is typical of the type of dog that is likely to develop separation problems, being very demanding of attention and becoming very attached to one person in the house and following that person everywhere. Purdy barked when the owner left her to go to work and the neighbours complained, so she was taken to work and left in the car. The first time she was left she chewed the gear stick and removed and chewed up both front seat-belts. She also did some damage to the front seats.

The owner sought our advice and because it was obviously a severe problem a full consultation was given and the necessary treatment plan drawn up. The owner worked steadily through this and eventually Purdy settled down, now being happy to be left at home while the owner goes out to work. By giving advice at the right time and helping the owner to understand the problem and realize that it would not last forever, we prevented the dog from being returned. The problem would have undoubtedly reoccurred in the next home, possibly to an even greater degree, and Purdy would have become one of the dogs that pass through many homes getting steadily worse with each change of environment.

In general, providing the solutions to behaviour problems can prevent many hastily returned adoptions and allows dogs to settle into their new homes until the bond between dog and owner has strengthened and the dog has learned how to fit in to the new way of life.

Repeated changes of home are very stressful for the dogs concerned and unless positive action is taken they have little chance of escaping a vicious circle of traumatic rehoming – the inevitable punishment administered by unknowledgeable owners in an attempt at a cure followed by return to the shelter for the cycle to begin again. A little timely advice to the owners can save the animal an ocean of suffering.

There is also an advantage to the shelter in that it only has to take in, care for and rehome a dog once and not many times.

Preventative measures

Since behaviour problems are one of the most common reasons for 'elective euthanasia' in adult dogs, socialization classes for 14–18-week-old puppies are run as a preventative measure in much the same way as vaccination is given to protect against disease. During socialization classes, early signs of problems, particularly those of dominance and nervousness can be noted and steps taken to prevent their further development. Lack of socialization is one of the main causes of fear-associated problems and these classes can do much to prevent them. It also gives an opportunity to educate owners so that they understand their dogs better, which leads to a better life for both dog and owner.

Puppies born at or abandoned at our shelter are socialized as much as possible from an early age. A supervised puppy playpen in our busy reception area, where puppies can be petted, played with and talked to (but not picked up), helps with this.

This is particularly important if the mother of the puppies has a nervous temperament and is likely to bark or show aggression to visitors looking round the kennels. It is likely that puppies of such bitches learn to be wary of or even aggressive towards strangers by watching their mother's behaviour during those impressionable early weeks and go on to exhibit aggression themselves as they reach maturity. Such bitches are housed in a place where the public have no access and the puppies' only contact with humans is in the playpen away from the bitch's influence.

Reducing the number of unwanted dogs

New owners invest a large amount of emotion when acquiring a new dog. Family members, particularly children, are understandably upset if they have to take a dog back to the shelter. This not only means that the dog then has to be rehomed once more, and uses up the resources of the shelter while waiting for new owners, but it is very likely that the previous owners will go out and buy a puppy rather than risk taking another dog from the shelter. This increases the demand for puppies and means that fewer unwanted adult dogs find homes.

Since there are so many unwanted dogs, organizations such as the RSPCA are forced to put healthy adults to sleep on a massive scale. Owners who continue to get rid of their dogs because they do not fit in with their way of life and then acquire new puppies, which often end up with the same behavioural difficulties, are only adding to this problem and increasing the number of animals that have to be destroyed.

Owners need to know that they can change their dog from a badly behaved one they want to give up into a well-behaved one they want to keep. Dogs should not be thought of as something to be disposed of or given away when they 'go wrong'. Our message to them should be 'don't get rid of the dog, get rid of the "problem"'.

The way forward

It is our intention to extend this work to the whole of the Blue Cross rehoming branches in due course, with the hope that other rescue societies will see the benefits and follow suit. Some rescue societies may say that they have too little time or not enough funds. We feel that it is essential to look at each dog as an individual, not as just one of many needing a home, and to make sure that each dog is rehomed successfully. Everyone involved in rescue owes it to the dogs that pass through their hands to learn as much as possible about dog behaviour therapy and to use it to educate and advise owners of problem dogs that will otherwise be given up.

The benefits to the welfare of the animals involved are obvious and we believe this approach represents a major step forward in the field of rehoming unwanted dogs.

9 Problems with People
Erica Peachey

The point of this chapter is to try to see things from the owner's point of view. It is written for those who deal with the dog-owning public and give, or attempt to give, advice which they hope will be followed. This includes dog training instructors, people working from veterinary surgeries, groomers, people involved with rescue work, pet shop owners, etc.

Our experience and knowledge of dogs means that we see things differently from the first-time pet owner. People can have a very strange way of looking at situations, and can be as difficult to understand as their dogs. This list of typical responses is not intended to be serious, a sense of humour is possibly the most important pre-requisite for working with the general public.

(a) 'He's never done that before,' *which means either* 'He always does that,' *or* 'I've never brushed him before so I wouldn't know.'
(b) 'He always does that,' *which means* 'I thought that as you worked with dogs, he wouldn't do this to you.'
(c) 'He doesn't like men,' *which means* 'He likes to bite men.'
(d) 'He understands every word I say,' *which means* 'Because I believe this, I don't have to bother teaching him anything, or make any attempt to understand him.'
(e) 'Castration? He'd never forgive me,' *which means* 'I don't like the idea.'
(f) 'He's not like that at home,' *which means* 'I would never try to examine his foot at home, because he would be just like that.'
(g) 'He's just friendly,' *which means* 'I have no control over him and cannot stop him leaping all over you.'
(h) 'He doesn't like to be left,' *which means* 'He wrecks the house if I am out of sight.'
(i) 'He's just talking,' *which means* 'I have never bothered to learn what he is saying.'

(j) 'He doesn't like dog food,' *which means* 'He has taught me to give him far more tasty food.'

(k) 'He knows he's done wrong, he looks guilty,' *which means* 'It justifies me losing my temper with him.'

In most situations, the dog is probably doing nothing other than behaving like a dog, whether he chews the furniture, messes on the carpet or tries to bite the postman, bites the children or savages another dog. Those of us who are involved with dogs would probably understand why the dog is behaving as he does and possibly would know the solution, but it is clear that the owner does not.

Is it fair to expect them to know? Until recently, very little information was available to an owner wishing to gain a better understanding of their dog. Even now, many people are simply not aware of why their dog behaves as he does. Many are not even aware that there is anything to understand. They haven't attended any talks, thought about it, discussed it and are not likely to read this book. They cannot be blamed for not knowing. Our aim should be to pass on the knowledge we have to the owner.

'It's always the people at fault, never the dog' and 'No bad dogs' are phrases which owners of difficult dogs hear all too often. This is not totally true as there are some very difficult dogs. The difficulties may be due to the breeding, the early environment or prior learning. Dogs from puppy farms can often present problems to unsuspecting people as can rescued or rehomed dogs. There are also difficult people who get it all wrong. Likewise, there are many caring owners and intelligent dogs who simply misunderstand each other.

Everyone makes mistakes with their dogs but most are fortunate and get away with it. None of us are perfect. However hard we try, humans cannot speak 'dog language' as fluently as even the slowest dog.

Some instructors fail to realize that the owner may have difficulty in acquiring new skills and knowledge. They seem to forget that it is not just the dog which may have difficulty in learning. Consider a few examples. Imagine how it would be if the driving instructor simply said, 'Well, I can make the car go forward, it will do it for me, why can't you do it?' He could show you many times how he can drive. However, until he explains how *you* can do it, you are unlikely to learn anything.

Beginners at horse riding are given a reliable horse to help them to learn. It would be dangerous if someone who had never ridden before took out a horse that had not been broken in. This is a situation where horse and rider have to learn from each other.

Guide dogs are trained extensively before they are introduced to their new owners and then a period of time is spent training the owner and dog together. It is not expected that everything will automatically be fine.

Is it therefore any wonder that problems develop between owners and their dogs? We need to define what a problem is. Generally speaking, it is something the owner wishes the dog didn't do, such as chewing the carpets or biting other dogs. Or maybe it's something the dog doesn't do which the owner feels he should, such as coming when called, or lying down quietly when visitors come. Such owners do realize that something is amiss, but others do not notice anti-social behaviour.

There are behaviours which we can identify as potential problems, which can and should be stopped before they cause any difficulties.

Imagine a pushy puppy in a dog training class. He is cute and sweet – and learning to be lethal. You show the doting owners how to stop him play biting and they say, 'We don't mind, it doesn't hurt.' You explain to them why the dog should be taught to accept being groomed, 'but all puppies do that,' they tell you, sounding slightly annoyed. You demonstrate how to get their puppy's attention and get him working with them, not against them, 'but we like him doing that,' they insist. When you ask them to control the games, they reply, 'he's only playing.' How do you get these owners to listen?

Picture a very obese dog at the veterinary surgery. It is obvious that her size is not helping her in any way, but any tactful suggestions meet with strong disapproval. 'She's big boned,' 'We're all big in our family,' or 'She's got a thick coat,' are some of the likely responses.

In another instance, a dog groomer is told by the owner, 'He doesn't like being brushed,' then has to attempt to re-educate the dog, and possibly the owner, so that the dog does not suffer in between visits. The owner does not want to know.

Once upon a time, there was a dog who was described to me as the 'typical dominant dog'. She belonged to a middle-aged couple with a grown-up family. They loved their dog.

'Anything she wants, she gets,' the husband told me proudly.

'Princess sleeps on our bed, she's our baby,' admitted the wife, smiling.

'We just love spoiling her,' he stated, as the dog leapt all over him and settled, briefly, on his shoulder.

The diagnosis was easy. My job was to persuade them. How could I motivate them to follow the advice I gave?

Only by explaining to them that their way of looking at things was not necessarily the way that their 'baby' was seeing it, could we hope to see an improvement. By showing them that she would be happier if she wasn't confused by their indulgences, they could begin to learn. By helping them to understand their dog, we were able to restore peace to the household.

Had I simply told them to stop her jumping on to the furniture, and to give her less attention, they would not have been able to do it. Had I told them that it was all their fault, that they had created the problems, they would not have even wanted to listen to me. If I had told them that I would not tolerate a dog behaving as Princess did, how would that have helped them? Yet all of these pieces of 'advice' had been given to them by well-meaning people before they saw me.

Even though dogs have lived with us since prehistoric times, it is not true to say that they are perfectly adapted to living with us. They are not human but dogs, a distant relative of the wolf, with instincts which have been adapted and modified, but which are still present.

You look at the animal and see a dog – wrong. You are actually looking at someone's pet. A pet dog is at least as important to that owner as your own dog is to you. The dog has his own personality and value to his owner and we must appreciate that fact, before we even consider working with them.

Anyone who has ever had any difficulties with any dog finds that the world becomes full of dog 'experts'. The next door neighbour knows what to do because she had a dog like that as a child; the man in the park tells you what to do because his dog did that until he 'showed it who was boss'; the lady in the shop can help because she has a niece who helps at the vet's surgery on a Saturday. And so it goes on. There is no shortage of people willing to give advice. However, there is a shortage of the right advice presented in the correct way.

Before we start trying to help owners and their dogs, we must first consider what it is they want from us. Until we know this, it is very difficult to be able to provide the advice they need in a way they will accept. We need to be able to see the situation from the owners' view point, and to understand how the owners perceive their dogs.

1. They must think you know what you are talking about. In other words, they must think you know about dogs in general. You know that you have gained a wealth of experience and knowledge, but why should people listen to you? You must show them that your knowledge can help them. A good reputation is gained through word of mouth, not by expensive advertising. Having chosen to ask your advice, they must have some faith in you. Your expertise is important, but most people will take it for granted and will expect far more.

2. They must think that you understand their dog. Not only must you seem to know about dogs in general, you must seem to know about their dog. This really impresses. Show them how well you know him by anticipating his behaviour, telling them how he reacts in certain situations. It is not fortune-telling or guess work. You should be able to obtain the answers from the dog, by observing his behaviour and body language.

107

3. Your must like their dog. All dogs have some good characteristics so compliment their dog – and mean it. You can comment on his eyes, coat, behaviour or anything else which comes to mind. If the dog is trying to bite you, you can admire his teeth! The comment *must* be genuine. If you can't find anything nice to say, look harder, there is always something.

4. Always be nice to the dog and the owner. A colleague of mine firmly believed that the way to stop a dog jumping up was to hold on to his paws. On one occasion, we visited a client of mine and were greeted at the door by the owner. The young dog jumped up to say hello, and my colleague grabbed the dog's paws and held on to them. The dog screamed, pulled away and hid in a corner. The owner, the dog and myself were horrified. The owner did not like what was happening, and had no control over what happened to her dog. This is not the place to discuss methods of teaching a dog not to jump up. Whatever methods you use, it is essential to explain your actions to the owner and ensure they understand what you are doing and are happy about it *before* you lay a finger on their dog. Ensure it is right for the dog *and* for the owner. If the owners ever have doubts about your methods of working, they will not listen to your advice.

Always wait for the dog to come to you. Be friendly to him, read his body language so that you do not make any mistakes. A surprising number of people feel that they have to 'make friends' with every dog, but not all dogs want this. Take it at the dog's pace, not yours. It is the owner who has asked you to help, the dog may not be so enthusiastic.

Be pleasant to the owner. This is easy with a responsible, caring, interested owner, but it is harder, and even more important, with an owner who appears not to be listening or to be raising objections. Speak to them, make eye contact, find some way to make them laugh and relax. The easiest way I have found is to bring their attention back to something positive about their dog.

Listen to the owner whatever views they may have. They know their dog, they care about him. They will always know things about their dog you cannot possibly know unless you ask them and listen to their answers. If the owner tells you that there is a bowl of food down constantly for the dog, you know why the dog is not interested in the titbit you have in your hand and that food is unlikely to be a useful reward. An owner will want to tell you what they have tried, and what effect this has had. If you listen, it saves you from being in the embarrassing position of recommending something which has already been tried and which has failed. It means you are in a far stronger position, because you can explain why that technique did not work, and why the methods you are outlining will.

5. They must know that you do not blame them. Judging owners is pointless. The necessity is to overcome this problem. You need to be able to explain in a positive and constructive way and you can't do that if you have preconceived perceptions and prejudices. So what if they have done things which did not suit this particular dog in this situation? We all make mistakes, but we should learn from them and adapt our methods in the future.

6. *You know how to help them.* Theory is fine, but all dogs are different and owners vary too, so you must be able to adapt your advice to suit their requirements and capabilities. A family with young children will have different needs and abilities from those of an elderly person living alone with only a dog for company. Treat people as individuals.

It is also essential to remember that the dog is used to being treated as an individual. About fifty per cent of dogs sleep in their owner's bedroom, and many of these sleep on the bed. Think of the implications of this, first, for the dog, who wants to be near to his owner, or his owner's bed; and second, for the owner, who possibly will not sleep without the dog being there. Bear this in mind when you expect a person to alter the relationship.

7. *They must know that you are on their side.* Owners want kindness, empathy and reassurance. They need to know that you are not going to write them off, tell them to do nasty things to their dog and that you understand their position. They want to tell you everything about him, his good and bad points. Although it may not be totally relevant, you must listen. They should know you appreciate their pet. They want to be sure that you will not hurt the dog (they have probably tried this already and found it didn't work), that you will not laugh at them (their friends will have done this), and that you can help them to improve the situation (as they are beginning to wonder if this is at all possible).

8. *Dealing with owners.* Instead of becoming angry or irritated when owners are difficult, look on it as a challenge. Problem dogs can make life interesting, and so too can 'problem owners'. Ask yourself 'why are they behaving like this?'

Is there a valid reason, such as they have tried many things already, and the dog is still getting worse? You must then explain why this has happened and why your suggestions will work.

Are they under stress? If so, make them relax, by saying something to make them laugh or perhaps talk about their previous pets or highlight this dog's good points.

Are they worried about their animal? They may be concerned that he will bite you, that you will hit him, or that you will tell them to have their dog euthanased. They have no idea of what you will recommend, so is there any wonder that they may be nervous?

There are many other reasons why we may consider owners difficult, so always try to see it from their point of view. Some time ago, I was attending a dog club where a man with a very bouncy Labrador pulled it around with one hand, with the other hand firmly in his coat pocket. I felt very annoyed, as I thought that he was not trying to control the dog. Then I found out that he had been involved in a car accident and only had the use of one hand. I had observed what was happening, but I had not bothered to stop and ask myself why he may be behaving like that.

The 'anthropomorphic owner' can be especially difficult. This is the sort of person who can only see their dog as a little furry person. The dog is denied any canine instincts or behaviours, and human values are forced upon him. This is the dog who 'understands every word I say' and who 'looks guilty' when he does something 'naughty'. Before any progress can be made, the owner must start to appreciate their dog for being a dog.

The 'macho owner' can be male or female. This type of owner feels he or she must 'dominate' their animal all of the time. This may take the form of physical abuse, which does not solve any problems, but does boost the flagging human ego. Again, until the attitude of the owner is improved, there is nothing more that can be done.

9. *They must want to do it and to have confidence that they can.* Motivate owners by explaining what will happen if they don't do anything, so that they will want to correct the problem. It is also essential to give them goals they can achieve. For instance, you may want them to play with their dog. For some owners, playing with their dogs each day is easy and enjoyable, but for others, it is difficult. This may be because they cannot get the toy back from the dog, the dog begins to be aggressive, or perhaps the owner cannot bend down or move quickly. Find out what they *can* do and build on this.

10. *You must show them how and show them it will work.* This is the practical side. We know that a dog will respond to a learned stimulus, but what does this mean to the pet owner? Usually nothing, so you must make it mean something. Show them that a titbit raised above the dog's head will make him sit, and show them how they can make use of this. It is not enough for a dog to lie down at the dog club when you ask him to. The owner needs to be able to teach the dog so that he will respond, not only at the club, but also at home and anywhere else.

11. *You must show them how again.* Studies show that people only remember a small percentage of information given to them. This percentage decreases if the person is at all stressed, is thinking of other things or if the words used are unfamiliar. However, we can increase what is learnt by repeating it, and by reminding people. Take time to ensure they understand. I am often surprised at owners who are given tablets by their veterinary surgeon, but have no idea as to the dosage or timing necessary. Of course they were told, but at the time they were thinking about their dog, and how to get him home quickly, and

so did not hear. We do not expect dogs to learn something perfectly after only one attempt, so why do we expect it from people?

12. You must help them to do it. Dogs learn and perform better when they are rewarded, and exactly the same applies to people (except that they don't respond so well to liver!). Increase their confidence and ensure they do things right by encouraging them. If you are not enthusiastic and positive, how can you expect them to be?

13. You must be there to help and support them. The harder something is, the more back-up people need. It may be hard because it is difficult for them, for example, when a person has to unlearn the habit of pushing and pulling his dog around. It may be hard because it is difficult for the dog, for example, when he has to overcome long established problems of aggression to other dogs. Either way, you must be there when needed to help, encourage, motivate and support.

14. You must be able to tell them what to do when things do not go according to plan. Questions starting with 'What if . . .' and 'Yes, but . . .' can be annoying, but they are actually a good sign. They show that the owner is considering what you have said and is trying to apply it to his own dog. It is far better for them to voice their objections so that you can give an explanation than for things to be left unsaid. Spend time overcoming any doubts that owners have. Tell them why your advice will work, and how to adapt it to their situation.

15. You must be prepared for excuses. The majority of excuses can be divided into three groups: lack of inclination, lack of time or lack of knowledge. You can help with all of these.

If they do not have enough inclination, you can motivate them by explaining what will happen if they don't do anything. Also, explain the benefits of doing something positive, and set small tasks with which they can be successful.

If they lack time, sympathize with them. You probably have the same difficulties. Give them things to do which will fit in with their life, rather than completely take it over. Very few of us have as much time as we would like to give our dogs, so we must make the best of what is available.

If owners do not have enough knowledge, give them more, in terms and language they can understand.

Occasionally, these comments are not excuses or doubts, they are actually shutters against what you say. They might as well say that they do not intend to listen to you at all. Find out how much motivation they have. If they have absolutely none, it is better to stop then, rather than waste your time and theirs. But it can be too easy to write people off in this way. Before you do, ensure that they really do not want your advice. If there is any doubt, you must continue to try to help, changing starting points if necessary. After all, why would they have approached you in the first place, if deep down they did not want some kind of help?

16. You must be genuinely pleased for them when they do it right. We also work for rewards, and it is very rewarding to observe a dog and owner progressing and understanding each other better. If you do not feel this, ask yourself why.

This chapter is not intended to be a definitive work on how to get the most from the public. It is intended to raise some points, and hopefully improve the skills that we are all constantly developing.

When working with people, you have to accept their points of view and, if necessary, use this as a starting point for your discussion. If one approach doesn't work, try a different one, until you find the required response. Then you can encourage and direct this.

Although you cannot treat every animal as you would your own, you can try to treat him as though he were as important to you as your own. Never underestimate how important a dog is to his owners. They are asking you about one of the most important parts of their life. They want some advice and they have chosen to ask you. Could there be a better starting point for success?

10 A Behavioural Approach to Training

John Fisher

Konrad Most's Approach

Training Dogs – A Manual was first written by (the then) Captain Konrad Most in 1910. He later became Colonel Konrad Most and his book was to become the recognized standard work on the subject throughout Europe. In fact the current Home Office manual on police dog training is largely based on Herr Most's theories and methods. A large percentage of pet dog training clubs in this country still rely very heavily on similar methods of training, even though there is a lot more knowledge available on training techniques and canine psychology.

His method treated the dog not as an intelligent human pupil imbued with a sense of duty but as an animal without moral values that learned not by logical thinking but solely through the faculty of memory. *Training by repetition became the standard*. This approach had a tremendous impact on the dog owning world.

Most also pointed out that the dog is a pack animal and he warned,

As in a pack of dogs, the order of hierarchy in a man and dog combination can only be established by physical force, that is by an actual struggle, in which the man is instantly victorious. Such a result can only be brought about by convincing the dog of the absolute physical superiority of the man, otherwise the dog will lead and the man will follow. If a dog shows the slightest sign of rebellion against his trainer or leader, the physical superiority of the man as leader of the pack must be given instant expression in the most unmistakable manner.

Physical punishment for incorrect responses became the standard.

Repetitive training exercises, using a praise and punishment regime, became the accepted method of achieving the role of leader of the pack: 'Me man, you dog, I say, you do.'

However, since Most's time tremendous changes have taken place in the way we live with dogs, not the least of which is that the majority of dog owners live in a more open-plan environment because central heating has now become the norm. Konrad Most was talking about service dogs who were generally kept in kennels when they were not training or working. The modern dog spends most of its time in whatever part of the house *it* chooses to occupy and often this will be the master bedroom. Konrad Most's dogs relied totally on their handlers for all of their comforts and privileges, feeding, exercising and most of all companionship. The modern dog is in general overfed, has a large garden to play in – usually with another dog which has been bought to keep him company – and chooses for himself which human member of the pack he will follow around.

In effect, the dogs of 1910 and before, right up to about the late 1960s and early 1970s were treated as dogs. They were not any the less loved than today, it is just that they were not granted the privileges reserved for the human pack members. They were fed after the family had eaten because a greater part of their food ration was made up of the family left-overs. Their freedom of movement around the house was restricted (a) because doors were kept shut to contain the heat of the coal fires and (b) because furniture and carpets were expected to last for years, people not being as affluent then as they are today. There were best rooms that were out of bounds to the family except on special occasions, and the dog was never allowed in them. As a pack animal, the dog understood that he was of lower rank, and because he didn't have the rank to object he would readily accept a Victorian-type training regime.

Konrad Most's training was designed to establish a pecking order at the same time as the dog learned the exercises. (To be fair, he was talking about service dogs.) In fact, the lifestyle of these dogs had already partly established this dominance/submissive relationship. It was his emphasis on man being the pack leader which revolutionized dog training at that time and still has a tremendous effect on it today. Of course, he was right to emphasize the fact that the dog is a pack animal, i.e. one which ranks each member of the pack (human or canine) in relation to itself. But his system of 'me master' training no longer works, simply because, over the last two or three decades, another revolution has taken place – a lifestyle revolution.

A pack leader has total freedom of movement, sleeps where he wants and always eats first to ensure he gets the richest pickings; it's nature's way of ensuring the survival of the fittest when food is scarce and instinctively indicates the animal's rank. Consider the position of the modern dog. Few dog owners now feed their dogs on the family's left-overs, in fact a lot of owners feed their dogs on the most expensive food that they can find, working on the principle that the dearer it is, the better it is. The modern dog now has access to all parts of the house, and if he is caught on the furniture it does not create a major family crisis. It is common for the dog to be fed before the family, so that after they themselves have eaten they will have nothing more to do whilst their favourite soap is on the television.

Any further comment would be superfluous.

Ranking
....

Contrary to Konrad Most's statement, dogs only very rarely engage in vicious fights to establish rank. Occasionally my APBC colleagues and I do deal with severe cases of sibling rivalry where two males (or females) from the same litter are genetically of equal ranking and ongoing increasingly violent disputes erupt. This is because in the home environment one dog cannot do what it would in the wild and leave the pack. Sometimes the same thing occurs when there are just too many dogs of equal rank in the house, or the owners are inadvertently promoting the rank of the wrong dog. In general, though, pecking orders are established surprisingly quickly and without violence.

The canine approach

Dogs instinctively know that if they are injured they cannot hunt, and that injuries to themselves or other pack members also reduces the pack's effectiveness as a hunting unit. Therefore, before they resort to violent confrontation between themselves, every peaceful avenue to establish rank is tried first. One only has to watch two strange dogs meeting in the local park to see how careful they are to avoid actual physical violence. They will stalk each other, they will adopt dominant stances, they might even display aggressive postures and vocalizations and sometimes even aggressive actions, but, providing we humans do not interfere, fights resulting in serious damage will rarely occur. Quite often all that will happen is that one will stand tall and the other will lower its head and avoid any eye contact. And that's it. Rank is established in seconds, then they get on with life.

The human approach

According to the teachings of Konrad Most, one of the primary aims of training is to establish the relative ranks of master and dog. Using the accepted training routines, this can take anything between eight and twelve weeks depending upon the particular club. Look how this violates the dog's basic instincts – we take weeks to establish what they do in seconds.

They first establish rank, then lay down the ground rules, most of which follow instinctively.

The most dominant dog will lead the way. We spend weeks trying to teach dogs to walk to heel and, until they learn, if they ever do, the dog does the leading.

The most dominant dog will precede the other through narrow openings like doorways. We spend weeks trying to teach dogs to do sit–stays and down–stays, and until they learn, if they ever do, the dog precedes us through doorways, gateways, narrow passages, up stairways, etc.

The most dominant dog will invariably win all competitive games, especially tug-of-war games for things like sticks. We spend weeks trying to teach them to retrieve articles which we throw for them and, until they learn, if they ever do, the dog wins these games of possession.

Certainly, what we are trying to teach them, not only makes living with the dog easier but will establish us as the higher ranking animal if we succeed. But all the time we are 'training' the dog, unless we are an expert at training, we are constantly undermining our own position by allowing the privileges of the highest rank, and if this is happening, we are wasting our time.

The old approach

The significant phrase in the previous section is 'unless we are an expert at training'. Obviously, the majority of people who attend training classes are not experts. The standard training regime of 'Tell the dog – make the dog – praise the dog' is very difficult to do effectively if you have not developed the split-second timing necessary to administer praise and correction when it is needed. Without this the dog is not likely to learn anything from the experience.

Although I accept the fact that in general dogs are not trained as regimentally as they used to be some years ago, drill-type training and repetition is still the norm in many clubs. I also accept that many dogs have been successfully trained using this format, but I wonder if perhaps they would have performed better under another system. The Konrad Most influence continues and is evident in many of the well-worn phrases that are still heard up and down the country.

- 'One of the functions of training is to enable you to dominate the dog.'
- 'If you can teach him to go down, you are teaching him to show you submission.'
- 'You should never use titbits, that's bribing the dog. They should do it because you have told them.'
- 'If you tell a dog to do something, you must be in a position to insist on instant obedience.'
- 'Dogs learn through a system of praise and punishment.'

- 'You cannot train a dog until it is at least six months old.'
- 'You must have the right equipment, in most cases a 'choke' or check chain.'

Those of us who have been seriously involved in the training of dogs have heard all or some of these statements many times. They reflect the standard attitude towards dog training. However much we soften the technique, for example, using the American technique of telling the dog to sit by tucking its legs inwards from behind the hocks with our forearms instead of forcing a dog's hindquarters to the ground with the flat of the hand over the root of the tail, or applying pressure with a finger to the kidney regions, we are still using the same 'Tell them – make them – praise them' principle, it's just less severe or softer in the approach. I deliberately used the word 'choke' in the last example of common phrases because this is what it was called when this piece of equipment was first introduced into the country. The name was softened to 'check' more recently, but in the hands of a novice owner/trainer, it is more likely to choke than check.

In effect, our methodology might be kinder and more pet owner friendly, but our basic principles have not altered.

A training or a behavioural approach?

Working as I do as a canine behaviour counsellor, I hear a good deal about the type of advice that is given to people about how to control the anti-social behaviour of their dogs. I am also the author and tutor of a correspondence course on understanding the human/canine relationship, and my students are always reporting the advice that is given to people with problem dogs. In most cases, this advice involves punishment for wrong behaviour without establishing its cause. A classic example of this concerns an eighteen-month-old German Shepherd Dog who had been referred to me after he had bitten his female owner during a training session. The circumstances were as follows.

Nero and his owner Anne were taking part in a group training class at a dog club. The exercise being taught was how to teach the dogs to sit. The owners were instructed to pull up on the lead with the right hand and at the same time to push down on the rear end with the left hand around the root of the tail. Nero struggled and leapt about every time Anne tried to do this. The instructor spotted that she was having problems and took hold of Nero's lead to demonstrate, but as soon as he tried Nero growled at him quite fiercely. The instructor lifted Nero clear of the ground by his lead and chain and shook him, at the same time shouting, 'No! No! No!' Nero urinated during this punishment. Anne was told to clean it up and when she had done so she was told to make Nero sit. When she tried, Nero bit her quite badly on the arm.

The instructor's advice was to have him put down as he was untrainable and had a vicious streak.

Anne reported the incident and the advice she had been given to her vet. Having known the dog since he was a pup, the vet was loathe to agree and suggested that she came to see me before making any decision. When I met Nero, he appeared to be a perfectly biddable dog of very good temperament and was obviously as attached to Anne as she was to him.

What was very noticeable, however, was the way he carried his tail. Instead of it drooping from the root as do most German Shepherds' tails, Nero's tail stuck out about two inches before it started to droop, almost like a docked tail. When I pointed this out to Anne, she said that he had always had a funny tail. Further conversation revealed that Nero rarely sat through choice, but he frequently laid down and licked his anal region and had gone through a spell of rubbing his bottom along the ground.

I at once suspected that there might be trouble with the anal glands and referred Nero back to their vet for medical examination after ensuring that there were no problems in any other area of their relationship. (Under normal circumstances the vet would have seen him first.) Nero had infected anal glands, which can be an extremely painful condition, and it was no wonder he did not want to sit. This was cleared up and Nero was taught to sit using the words 'park it'. Anne had to change the command because one of the side effects of the traumatic experience at the dog club was that Nero would run away when he was told to sit.

There are many stories of 'training' cures being tried on what has turned out to be a symptom of some other problem. What annoyed me about this case was that the signs of his condition were so strikingly obvious. His reluctance to be placed in a sit position should have rung some warning bells in the instructor's mind. Instead, he steamed in to *make* the dog obey instead of first asking *why* the dog was not obeying. Nero could have been unnecessarily euthanased for being an aggressive dog.

As we will see in the next section, there is a way of training dogs without resorting to punishment. As John Rogerson (APBC) puts it, the difference between a traditional trainer and a behavioural trainer is that the first controls the body whilst the other tries to establish why the body has gone out of control.

I am firmly of the opinion that anyone who offers advice to pet owners about their dogs' training should have a reasonable knowledge of behaviour (not necessarily expert), an ability to suspect some medical involvement (no more than they would do with their own dogs) and a constant question in the forefront of their minds, *Why is this dog rejecting my training methods?* The difficulty with Nero would never have gone as far as it did if this approach had been used.

The behavioural approach can pinpoint a possible medical problem but not diagnose it. And it helps that a practising behaviourist or a dog club instructor has more time available to them than the consulting vet is allowed. If we can establish that changes in the dog's normal behaviour correlate with changes in eating, drinking, defecating, urinating, normal activity, coat colour or increased hair loss etc., it is worth reporting this to the vet because it might be a sign that there is some medical reason for the behaviour. If the problem is coincidental with changes in the environment, then the vet might be prompted into referring the case on to a practising behaviourist. In most cases, the forward thinking club instructor is capable of advising some effective means of behaviour modification, but we should all be aware of our limitations and be prepared to say, 'I don't know, but I suspect' and then suggest that the owner consult their vet for further advice. Similarly, vets should be aware of their own limitations, especially in the field of training, and accept the reports as an aid to their diagnostic abilities. Veterinary care, behaviour counselling and good training advice should go hand in hand for the benefit of the man–dog relationship and it is time that we all realized this fact.

It is a peculiarly human trait to want to punish 'bad behaviour', which in most cases turns out to be anything that goes against our particular human standards. This attitude allows no exceptions, even when the offender is a dog and is doing what is natural to it – barking, growling or biting.

I think the first, and most difficult obstacle for us to overcome is our deep-seated need to express our anger and extract some sort of revenge. If a dog defecates on the kitchen floor overnight and another dog comes in in the morning and treads in the mess, does it attack the culprit for what it has done? It does not, but the human does. Are we forward thinking enough, as we look at the mess squeezing out between our toes, to ask why we put this dog into such a position that it couldn't help soiling its own den? Or do we immediately take the view that it has been a bad dog and needs to be punished in order to teach it not to do it again?

Before we look at how dogs should be trained to perform what are basically 'tricks' – Sit, Down, Heel, Come, Fetch, etc. – we should really be sure how we would personally handle such an event. Dogs cannot share our social standards, and to attempt to teach them human values is a futile exercise. Conversely, humans have the ability to learn about canine values and to, as it were, outwit the dog at its own game. However, as I said earlier in this chapter, unfortunately our modern lifestyle favours their efforts to gain rank more than it does our attempts to lead through training, and as a result the traditional approach is no longer as effective as it used to be. We are losing the right to give them commands and are therefore in danger of allowing dogs to become a nuisance in our society. This problem has to be addressed and corrected before we start to teach them what we mean by the words of command, and the advice given in many of the other chapters will enable you to do this.

A behavioural approach to training

I have a standard answer to a question I am asked on numerous occasions. The question is 'How do dogs learn?' The reply, 'Very easily, if we use the right approach and apply common sense.' The old techniques will work to a certain extent if we first establish the rank, but if rank has already been established, why run the risk of creating resentment and mistrust by physical manipulation and punishment for getting it wrong? Why not teach the dog to enjoy doing what we want it to do?

The basic principle behind the behavioural training approach is simple. If the dog does something and finds it rewarding, the chances are high it will do it again, and if it doesn't find it rewarding, it is unlikely to do it again. Therefore, the techniques used in behaviour modification programmes – the correct application of positive and negative reinforcements which coincide with the desired or undesired behaviour – can be used to teach a dog anything we want it to learn. Before we go any further we should first of all examine what the difference is between positive/negative reinforcement and reward/ punishment.

Reinforcement is what happens during the act, whilst reward/ punishment happens after the act. Ideally, the dog should only perceive its own actions as being rewarding or not, and not as something we get it to do before administering a reward or punishment. Let me give you an example of negative reinforcement as against punishment.

In my office there is a calor gas fire. The modern dog is generally unaware of the heat which can be generated by a naked flame. If the fire is on, it is usually one of the first things that dogs investigate. They approach cautiously (all dog owners have seen the long-necked investigative approach towards a new stimulus) but before any real harm is caused they sneeze and back off. They have just learnt not to go too near the fire, but they cannot blame anyone else for the discomfort. That is a negative reinforcement which has an instantaneous learning effect without ruining the relationship between dog and owner. *It was his stupid fault!*

Alternatively, I could take hold of his collar and put his nose near the flame until he struggled to get away. The learning effect is just the same (the fire burns) but the danger is that our relationship will be ruined because we were directly involved with the negative aspect. *It was our stupid fault.*

Now let's look at an example of positive reinforcement as opposed to reward.

The seal has failed on your fridge door and the smell from your steak is leaking out. The dog investigates this smell and scratches at the door – no result. He barks at the door – no result. He sticks his nose at the point of the strongest scent and sniffs – no result. Because the sniffing gave him more satisfaction than the other behaviours, he tries it again but this time more forcefully – the door bursts open and he gets the steak. *He has just learnt how to open the fridge.*

Alternatively, to teach him to open the fridge, we could sit him in front of it and play a tug-of-war game with a toy (letting him win the toy as a reward). We could then tie the toy to the fridge handle and encourage him to pull it, giving him an alternative toy or a titbit when the fridge door opened. In this way we are rewarding the right behaviour, but how long do you think it will take the dog to realize why he is getting rewarded? In most cases he will soon realize that when the toy is tied to the handle, he never wins it and therefore all of his concentration will be on the alternative reward toy to the detriment of what we are really trying to teach. At the end of the day, or week, or month the reward-based technique will eventually work. However, there are too many confusing factors involved and these will be detrimental to the dog's ability to learn quickly and efficiently.

I am sure that APBC member Dr Roger Abrantes will forgive me for repeating his brilliant explanation of how reward-based training can be wrongly applied. He said that in his country (Denmark) the most popular toy reward for a dog is a ball. What he sees in a lot of clubs is that they torture a dog for fifteen minutes and then they throw him a ball.

I know he didn't mean that they physically torture the dog (maybe some trainers do) but this simple statement sums up the difference between reward for performing the act (reinforcement reward) and reward after the act (unconnected reward).

It is vitally important that we establish in our own minds the difference between reinforcement and reward, or negative reinforcement and punishment before we further investigate a behavioural approach. The two methods are distinctly different, but the phraseology is very similar. When I talk about 'reward' in this chapter, I will be talking about an incentive that the dog recognizes is on offer but has to work out for itself the best way to get it, or something that the dog does of its own free will which subsequently brings it reward. There are a lot of trainers who now say that they use a reward-based system of training, but in effect they only substitute favourite toys or yummy titbits for what used to be verbal praise. In most cases, they are still using – admittedly more restrained and kinder – physically orientated methods to get the dog to perform the required movement in the first place. This reduction in force and increased reward is encouraging, but it still does not constitute a behavioural approach. Physical manipulation into a sit position for a pat on the head, a stroke on the chest, a titbit or thrown toy is a reward-based programme. Presenting an incentive (without saying a word) and then waiting for the dog to experiment with a variety of postures and actions until it discovers that sitting gets rewarded is a reinforcement-based programme. If you think about it, the punishment then becomes not getting the reward – but that's for the dog to decide, not us.

Waiting for the dog to do what we want it to do without getting annoyed because it is not instantly obeying our wishes is a hang-up entirely peculiar to humans. Our impatience is a reflection of over ten thousand years of a very close relationship, with *us* having the attitude that we are top dog. We have been involved with other species for a very long time, but how have we learnt to train them?

How would you get an elephant to sit?

How would you get a rat to learn a maze pattern?

How could you be totally confident that in a sea life demonstration in front of an audience of hundreds the porpoise would play football and the killer whale do a double somersault ten feet clear of the water?

We have learnt to train them through reinforcement, and we have been able to do this because we have never been able to identify with them as a similar species and have always accepted that they have different values. If it is rewarding, they will do it again. If we call the cat in from the garden and it doesn't come we shrug our shoulders and accept that cats are cats. If the dog doesn't come we get angry and start to plot revenge. Our problem is that we have always regarded the dog as a species just above animal and just below human. In fact all animals, including humans, learn in the same way.

The way forward

If we can establish a clear-cut dominance/submissive relationship between us and the dog, then all we need to do is to teach the dog what we mean by certain words of command. It is important at this stage that we correctly understand the difference between dominant and submissive – perhaps better described as higher and lower ranking. In all animal societies, this is established through respect and trust and not through strength and fear. A good relationship is built on a genuine preference for the particular company of another two-legged or four-legged animal, and the same is true in reverse. Companionships of this kind are based on emotional attachments and common interests, not on strength versus weakness.

Therefore there must be a genuine bond between you and the dog as part of the family unit, not based on a particular role that the dog has been obtained to fulfil. Some common examples of this are the dog that is obtained as a guard dog, the hobby dog that will further one's own ambition in competition, the dog the kids wanted, the dog that is an extension of one's own ego because of the originality of the breed or that will enable one to keep up with the Joneses, and so on. Dogs are extremely sociable animals and need to feel part of a unit, but they must know their place within this unit.

In most pet homes, this genuine regard for the dog is without question, but the way the dog is treated tells it it is of a far higher rank than it is equipped to cope with. In effect, we need to love and respect our pet dog as part of our family unit, but still ensure that it knows its place within our pack structure. In short, we must attain the right to give commands, and John Rogerson's chapter gives some positive help in this area if things have gone wrong.

We are now left with just the mechanics of teaching a particular movement following a particular verbal or visual cue, and this is incredibly easy to teach. When I lecture on this subject to dog trainers, the general comment is that I make it sound so easy. There is a reason for this – it is!

How many of you acknowledge the response you get from your dog when you pick up its lead, or have got to the stage where you have to spell the words w-a-l-k or c-a-t-s because the dog reacts immediately? I'm sure that many of you who started doing this are now convinced that your dog can spell!

What does your dog do when the *News at Ten* theme tune is played on the television, either the first time (if you take it out at the start) or the fourth time, because of the ads in the middle (if you take it out at the end)?

How many of you are prompted into feeding your dogs because of their behaviour at a particular time of day and wonder how they know the time of day?

How many of you find that your dog does not always respond to the recall command 'come' but will always respond to 'whowantsabiccy'?

I defy anyone reading this chapter and who has a dog, or who has had a dog in the past, to say that there was any word or action that did not guarantee an immediate positive response. Because these learned responses do not come under our 'human view' of training, we tolerate them and often encourage them to become party pieces. But the long-term learning effect of this regular occurrence – no pressure, just reward – affects the dog's behaviour for life. Let me give you an example from my personal files.

Before I met my wife Liz she had a Weimaraner puppy called Oliver. She used to take him out with her then young children Mark and Jo. The children were instructed to leap in the air and whoop with joy if Oliver urinated or defecated during a walk. Oliver died when he was ten and despite the fact that the whooping and jumping had ceased years ago, he always displayed stupid puppy behaviour and always did a lap of honour whenever he defecated, even if there was no one there at the time.

So what conclusions can we draw from these examples? If the reward is good enough, the dog will learn from it and will do it again. If it is used to being rewarded for its behaviour and the reward stops, it will try harder the next time to earn the reward – it becomes a conditioned response. If we can overcome our indoctrinated belief that a dog must do something because we have told it to do so, and accept the fact that dogs can be taught to do things with incredibly reliable accuracy, providing the incentive is good enough, then we can teach a dog to do anything we want it to do – providing it sees that we have the right to tell it when the alternative is more rewarding.

It is not within the scope of this book to go through each dog training exercise and describe how to teach it using a reward-based technique, but my recent book *Dogwise: The Natural Way to Train Your Dog* does this in detail. The purpose of this chapter is to get you to understand the theory behind the technique and if you do this the application becomes a matter of common sense. Usually the quickest way to teach an exercise is also the simplest, and the only thing which stands in the way of our realizing it is our so-called superior human intellect. If I use the recall exercise as an example, you will see what I mean.

The object of the exercise is to teach the dog to come and sit in front of its owner. When I talk to training groups, and after I have explained the theory of behavioural training, I hand a few of the group a titbit, point to some dogs on the other side of the room and ask them to demonstrate how to teach a recall using a food reward.

Some encourage the dog to come to them by catching their attention, giving the 'come' command and then running backwards.

Some prefer to use a lead and give the dog a gentle tug with the command and run backwards.

Some try and get the dog to stay, take a few paces back and call the dog to them. If they have difficulty with the 'stay' they ask someone else to hold the dog and in every case, they go a lot further back because they are confident that the dog cannot follow until it is released.

Some get down on their hands and knees to present a less dominant posture and give the dog more confidence to approach.

Regardless of the technique used, the principle is the same in every case – to get the dog to approach them to receive its reward. It is after all a *recall* exercise.

I then demonstrate an alternative method. I approach each dog and stand in front of it in the traditional 'present' position. I show it a titbit without allowing it to eat it. I then move the titbit slowly above its head and slightly behind the eyeline so that it has to look up and backwards.

A dog's skeletal structure is designed in such a way that its rear end has to go down to enable its head to go up and look behind. In effect, it has to sit to see the titbit and as it does I say, 'Come.' Generally, this is received in total silence by the trainers.

At first I found this silence disconcerting, but now I find it amusing. They have spent years trying to teach dogs to come forward and sit in front of them, and now I suggest that all they have to teach the dog to do is to sit in front of them on the command 'come'. If the dog learns this concept, it doesn't matter where it is or what it is doing when it hears this command, it has to get to you for its reward. The result is a recall, and the same technique works for the retrieve.

The silence results from the shock of having ten, twenty or even thirty years of experience in how to cure recall problems overthrown in a moment. Their first reaction to learning that they can teach the bottom line and then ask the dog to work out how to reach that line is that they have been wasting their time all those years. This is not the case, of course. They have gained a deep expertise of timing, the ability to read a dog and the knowledge of alternative methods of training for every exercise which is priceless. What I offer them is simply another tool for their toolbox, one that is simple in design, easy to operate and effective for the job in hand, as they find out when they return to their usual training environments.

I then go on to suggest that teaching a retrieve only requires the dog to learn to sit in front of you with something in its mouth when you say 'fetch', and that it will only receive a reward if it sits still and doesn't chew the article. If the dog is only rewarded for sitting straight and calmly holding something in its mouth when it hears the command 'fetch', then it has to go out and pick up whatever you have thrown to enable it to do this. So simple, but, like the recall, so effective.

It is in the area of the 'down' command that the difference between the traditional and behavioural approaches starts to raise questions. It is commonly said that the 'down' is a submissive posture for a dog (not always the case when you see dogs displaying active submission). This long-held attitude towards the down position has resulted in physically firm methods of teaching accompanied by harsh, insistent commands. A 'rapid down' is also taught so that a dog can be stopped instantly if it is running towards a road or some danger. Teaching this technique involves even greater physical firmness and louder, harsher commands.

This approach dates from the advice given by Konrad Most, who said that to get the dog to lie down at a distance from the trainer it is necessary to resort to a strict form of compulsion. He then went on to describe a technique which involves accompanying the compulsive command *'down'* with a sudden, very swift cut from a switch across the dog's back. I have seen modern trainers using the same technique, but hitting the dog with its lead. They excuse their actions to the concerned pet owner by saying that this will save the dog's life when it is in danger.

Pushing the dog to the ground, pulling it to the ground with the lead under the trainer's foot, pulling its legs out from underneath it are all commonly used compulsive training methods. They are designed to achieve the desired result, while convincing the dog of the trainer's physical superiority. Roger Abrantes once told me of a technique he witnessed in Germany where the class under instruction were taught to kick the dog's legs away from behind. As he ironically commented, the theory is sound, without the legs the body will go to the ground.

However, as this chapter has emphasized pack superiority should be established before formal training starts. Rank established, we need only teach what we mean by the words and this teaching process is simple, not only to teach the 'down', but the 'sit' and the 'stand' as well.

Progressing each exercise (shaping)

Once the dog will adopt the required position when you offer the reward, the movement can be put under command control. You would not tell a dog to 'sit', for example, until you are sure it will do so when you hold a reward above its head. If it jumped up and snatched the reward, it would learn that this is what 'sit' means. Command control established, you now ensure that you get a first command response on a ten out of ten basis. The exercise can now be progressed and to do this we employ a technique known as *shaping*.

Shaping involves withholding the reward until a bigger, better, or faster response is attained. Having learnt that a particular action is rewarded, the dog tries a bit harder when no reward is given. This improved performance then earns the reward, but only if he maintains the improvement – anything less is ignored. The important thing to remember about shaping behaviour in this way is not to make each step too difficult, and never to reduce the criteria for getting a reward once an improved performance has been achieved.

Using these techniques, the retrieve, for example, might be taught in the following way.

Step one. Each time the dog 'mouths' the offered article he gets a tit-bit. When he mouths it as soon as you offer it ten times out of ten, you *raise* the criteria for reward.

Step two. A reward will now only be given when the dog grasps the article in his teeth, so you withhold the reward until he does so. Initially, he will mouth and nudge it with increasing firmness, but you just ignore this. Very quickly he will attempt to grasp it, which is rewarded. You now wait for the ten out of ten level of understanding.

Step three. He now has to grasp it and pull it out of your hands. You withhold the reward and wait for him to find out what he must do to earn it.

Step four. He now has to grasp it, pull it out of your hands and hold it to a silent count of three.

And so it goes on, little steps leading up to him sitting in front of you with the article in his mouth. Now you say 'fetch'. If at any stage you have difficulty in progressing, it is a good idea, when you finally get the desired action, to increase the reward. Change to a tastier tit-bit, increase the verbal praise and enthusiasm on your part, invite the neighbours in for a party – really let the dog know how pleased you are. This is called a 'jackpot reinforcement' and is no different from us working for a weekly wage but occasionally getting an unexpected bonus. It increases our output and willingness to do better.

This type of training should be of just a few minutes' duration and be spread over a number of days. Each dog will learn at a different pace. Once the first few steps have been taken, the whole process snowballs at an incredible rate, but the time needed to establish the principles of the contract between you and your dog will vary with age, breed and owner. Having said that, *all* dogs, regardless of these factors, will learn using this positive reinforcement–random reinforcement–jackpot reinforcement method.

No mention has yet been made about teaching the dog to walk to heel, which is generally the first problem addressed in any formal training programme, and where, in my opinion the biggest mistakes are made. If your dog regards you as the pack leader, responds to your requests to Come, Sit, Down, Stand, Fetch, etc., it will invariably not pull on the lead. In effect, pulling is a symptom of the dog's dominant attitude. Trying to overcome a symptom without looking at the root cause is a folly.

It is an absolute certainty that a dog will pull on the lead if the owner cannot control it when visitors arrive; if they have difficulty in getting the lead on in the first place because the dog is so excited about going for a walk; if they cannot stop it barking at the slightest sound; if it whines and paces in the car; if it growls at them when they approach its food bowl; if they cannot speak on the telephone without the dog badly needing a pee, or pinching one of their valuable belongings, or demanding to be stroked. Pulling on the lead is a natural extension of their leader attitude, which all these behaviours demonstrate.

When I look back through my records of veterinary-referred problem dogs, pulling on the lead (despite previous formal training programmes) stands out as a reported secondary problem in dogs which were brought in with a primary dominance-related problem, such as aggression to strangers, aggression to family members, aggression to other dogs, hyperactive attention seeking problems, food and toy guarding behaviours, chasing joggers, sheep, horses, bikes and cars etc. Dogs which were brought in with problems that were not dominance-related – fearfulness, phobias, destructiveness, house soiling, howling and barking etc. – did not pull on the lead. It seems logical to assume therefore that pulling on the lead is more of an attitude problem than a training one. Whatever the reason, if a dog has pulled for months, and sometimes years, there will be a learned element to overcome.

It should be understood that there is a difference between the exercise 'heelwork' and the exercise 'don't pull'. Using the type of behavioural approach already described, most dogs can be taught to walk to heel without using a lead to influence their behaviour. Stage one would be standing still with your dog alongside your left leg (not sitting) waiting to be rewarded. Stage two would be to move two steps before giving the reward; then perhaps five steps and a half turn etc. I do not need to repeat the whole procedure, it's just a basic reward-based (shaped) exercise. The dog can be taught not to pull on the lead in the same way, unless of course it has been allowed to form a deep-seated habit on the matter.

There is a very simple explanation as to why dogs pull on the lead – it is because there is something to pull against. Regardless of whether the dog is pulling forward to get everywhere first even though he does not know where he might be going; pulling to one side to sniff at every vertical surface (the famous Collie lean); or digs his paws in when he finds something interesting and refuses to budge – in all these instances the dog is resisting the pressure it feels from the lead. Take away this pressure and there is nothing to resist. Sounds simple, doesn't it? That's because it is simple, like all the techniques I have described so far.

In my previous books, *Think Dog, Why Does My Dog?* and *Dogwise*, I describe a technique that uses the ratchet of an extending lead to interrupt the dog's attempts to pull. The basic principles behind the effectiveness of this technique are:

A. By not applying the brake there is no pressure for the dog to resist and therefore nothing to pull against.

B. All that can happen is that the dog will walk faster than the owner, but a gentle application of the brake when it tries to do so will set up a vibration within the lead that will interrupt their leg movement.

The majority of dogs quickly learn that it is a futile exercise to try and walk at any pace other than the one decided by the owner. Although it is not a technique which requires any strength, it does require a certain degree of timing and expertise, and I always advise people to practise it with the lead attached to something solid before they try it on their dog. The benefits to the dogs are that they can be taught on an ordinary buckle collar and they don't get their heads yanked off in an attempt to teach them not to pull. This method uses the same negative reinforcement principles described earlier – pull if you want to but you will find it's not rewarding; not pulling will be rewarded. It's simple, effective and kind.

The way forward is to reject punishment as a training aid and adopt a more positive approach. The other authors involved in writing this book will help us all to understand our dogs a bit better. If we can do that and develop the right relationship between us and them, then teaching them to respond to our commands is very simple if we use the techniques which I have described. As I have explained, the way we live with dogs has changed and the way we train them should change also. Considering some of the techniques still being used which I hear about all too often, it is a tribute to the temperament of dogs that so many of them allow us to do so.

11   Behaviour Problems in the Cat

Peter Neville

The nature of the cat–human relationship

Cats are different from other larger companion animal species such as the dog in that they do not relate to their owners as part of an organized social or hunting group. The cat retains its ability to hunt as a solitary animal and continues to do so away from the shared human den, even when well fed and cared for. It also retains a high level of adaptability and can survive in a wide range of environments from sub-antarctic islands to semi-desert as a free-living creature deriving little or no benefit from man. Indeed, recent evidence from research in South Africa suggests that what is nowadays termed the domestic cat varies genetically very little from its wild ancestor, the African Wild Cat. The cat as a pet is apparently simply occupying a favourable niche in our homes without compromising its abilities to survive, reproduce and exploit almost any other opportunity in the world where an adequate year-round supply of food and water is available.

The cat has succeeded in establishing itself in our homes because it is willing to engage in friendly and affectionate behaviour with us, despite not being dependent on such social relations. Our pet cats' predatory or territorial aggression is usually restricted to the surrounding gardens and they happily return to our homes for rest, affection and security. The human–pet cat relationship is composed of many contradictory elements. Indoors the cat is valued for its cleanliness, affection and highly evolved play behaviour, for which it is much admired. Although still essentially a wild creature, and not a group hunter, the cat has an enormous capacity to be sociable and can accept the benefits of living in the human family and den without compromising its general self-determining and independent behaviour. Outdoors, it is clear that this sociability should in no way be confused with domestication, for it continues to be able and to desire to hunt, even when well fed.

The cat views its human family partly as social conspecifics and partly as maternal figures, continuing much of its kitten behaviour into adulthood when with them. This frequent demonstration of affectionate responses of an infantile character helps to build an extremely strong bond between owner and cat, one that is essential to bear in mind when treating medical or behavioural problems. When their pets present behaviour problems, most cat owners will tolerate a much higher level of disruption to their social life and household hygiene than will dog owners, and are far less likely to apportion blame to the cat. Cat owners are often very sensitive to their cat's emotions and accept that they may vary from day to day or even hour to hour. Even though some breed characteristics of temperament have been coincidentally enhanced in breeding and selection for physical appearance, for example, docility in longhairs or responsiveness and attention demanding in the Burmese, the cat's basic character is largely unaffected by domestication. Most owners understand that pet cats are not easily trained to perform set tasks, and so accept that their cat may not be causing difficulties deliberately when behaviour problems arise.

Behaviour problems

A breakdown of the cases referred to me in 1990/91 by veterinary practitioners for treatment at clinics held at veterinary hospitals throughout the UK and at the Department of Veterinary Medicine at Bristol Veterinary School gives some interesting results (see diagram on page 132). Pro rata, owners of high-value breeds were more likely to seek help than owners of crossbred cats. While only 8 per cent of British cats are of recognized breeds, 44 per cent of the cases involved pedigree strains, 14 per cent first-cross pedigree strains and 42 per cent domestic short- and longhairs. Of the pedigrees, 24 per cent were Siamese, chiefly referred for problems of indoor urine spraying; 20 per cent Burmese, chiefly for problems of aggression towards other house cats, or, if allowed access to the outdoors, towards local rival cats; 13 per cent Abyssinian, mainly for sudden breakdowns in relations between several Abyssinians sharing a home and 13 per cent Persians, almost exclusively for serious house-training problems. Representatives from seventeen breeds were referred. Just over half the cases referred (56 per cent) were male and only fourteen un-neutered cats were treated. The average age of cats presented was three years and ten months, though this is somewhat skewed by the presence of five cats over fourteen years of age treated for loss of house-training and excessive night-time vocalization. Most of the cats seen were between one and five years old divided as 1–2 years (26 per cent), 2–3 years (21 per cent), 3–4 years (18 per cent) and 4–5 years (11 per cent). The most common case profile is therefore a neutered 1–2-year-old male domestic short-hair or Siamese cat which lives with one

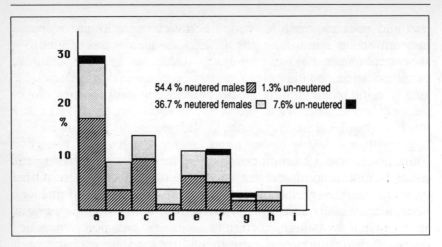

a *Indoor spraying*   b *Other indoor marking (scratching, urination, middening)*
c *House-training (loss of)*   d *Nervous urination*
e *Nervous conditions, e.g. fear of visitors, agrophobia*   f *Aggression to other cats*
g *Aggression to people*   h *Self-mutilation*   i *Other*

other cat and which sprays or soils inappropriately indoors. The implication of this finding may be that would-be owners should avoid keeping such types of cat, but the majority of cats, even of this type, do not present behaviour problems and make excellent trouble-free pets. In any case, there is no firm evidence to suggest as yet, as there is with dogs, that many such problems could be avoided with improved breeding, character assessment or selection of the pet when young. Some kittens simply turn out to be more reactive, sensitive or problematic adult cats despite careful breeding, the best of husbandry and attention to social development.

The emphasis on referral of pedigree strains may be because an owner will often have paid a relatively large sum of money for their cat and so be less willing to reject it because of a behaviour problem than if they owned a crossbred cat which would be cheaper to replace. Pedigree strains are also more likely to be housed permanently indoors (and so more likely to present noticeable problems due to being more reactive to change within the home) or because the most popular breeds such as Siamese, Burmese and Persians are often reported by owners and breeders as being generally more 'sensitive' or emotional.

Very few feline behaviour problems have simple answers and, in diagnosis and treatment, most necessitate the detailed recording of a complete problem history, relevant medical history, lifestyle and relationship data within the family. The nature of the home environment is a crucial cause of many problems and, through modification of access, in the treatment of most. The treatment of behaviour problems only with drugs of any kind is rarely curative and usually inhibits learning. The use of drugs often has to be based on a trial and error approach as there are marked variations in response between breeds and individuals. Where used, drugs are offered as a short-term measure to facilitate the application and acceptance by the cat of management, husbandry or behaviour modification techniques and to facilitate learning by degree during systematic desensitization.

Indoor marking (scratching, spraying, middening)

Chin, head and flank rubbing are normal forms of scent marking and social communication and are encouraged by most owners. Other types of marking behaviour may not be well tolerated indoors. Scratching to strop claws can usually be transferred from furniture to an acceptable sisal-wrapped post, hessian- or bark-faced board by placing this in front of the furniture and then steadily moving it to a more convenient location. Scratching as a marking behaviour is usually more widespread in the home and appears to be performed as a dominance gesture in the presence of other house cats. It should be treated in the same way as other forms of marking, such as urine spraying, associative urination and defecation away from the litter tray. The latter two actions are usually performed on beds or chairs, or sometimes even on or in hi-fi headphones where the owner's smell is most concentrated. Presumably the cat does it because it perceives a benefit in associating its smell with that of a protecting influence against challenges, real or imagined. Most commonly owners report toileting on the bed when they go on holiday and leave friends or minders to care for their cats. Associative marking can also occur on doormats, where challenging smells may be brought in on the owner's shoes from the outside, and, unusually, on electrical appliances such as video and washing machines, perhaps because of a scent association.

In contrast, urine spraying is a more normal and frequent act of marking practised outdoors by most male and female cats, entire and neutered. They spray in a standing position and usually direct a small volume of urine backwards against vertical posts, such as chair legs, curtains, etc. Cats usually have no need to spray indoors because their lair is already perceived as secure and requires no further endorsement. It has been suggested that those cats which spray indoors may also be more restless generally and more active nocturnally than non-spraying cats. They may also be relatively more aggressive towards the owner.

Of the indoor spraying cases treated by me in 1990/91 (see diagram on page 132), 47 per cent came from two-cat households; only 10 per cent were solo house cats and 17 per cent came from three- or four-cat households. Most were male neuters (61.5 per cent) indicating perhaps some social inability of males especially to accept the presence of another cat or cats in the home territory up to some individual threshold beyond which there may be some suppressing effect on the need to spray. This contrasts with the previously generally held view that the greater the number of cats sharing a house, the more likely that one at least would spray. I have visited one three-bedroom home in London used as a rescue centre for stray and unwanted cats that houses over 140 resident neutered cats in addition to a continuous input of new arrivals and departures when individuals are found new homes. The owner reports that none of her cats has ever sprayed, and her home does indeed smell relatively clean! Other cats seem peculiarly specific in the objects that they spray in the home. One famous sprayer only ever anointed a picture of the Czar and Czarina of Russia hanging on the stairway wall, a feat achieved by the cat by reversing his rear end through the landing balustrade and targeting a jet of spray some three feet across the stairs!

When such marking occurs indoors it is usually a sign that the cat's lair is under some challenge. The challenge may be obvious: for instance, the arrival of another cat, a neighbour's small dog, a new baby, an increased challenge from a cat outdoors, or it may result from moving or changing furniture, redecorating, family bereavement, having guests to stay, bringing outdoor objects anointed by other cats into the house, bringing in novel objects (especially plastic bags) and, most commonly of all, the installation of a cat flap. This can totally destroy indoor security for a cat, even without the obvious challenge of a rival cat or a neighbour's small dog entering the inner sanctum of the home. Spraying by at least one cat may be more likely where increasing numbers are expected to share a home base, as described above. Some highly manipulative and social individuals of certain oriental breeds engage in protest spraying when frustrated or denied the attention of the owner.

Treatment

The function of spraying and other marking is not just to leave messages or challenges for other cats, even though feral cats are clearly able to distinguish between urine marks deposited by males, females, and neutered or entire individuals. The response of other cats to a spray mark is often to investigate it but then to continue normally rather than run away or show a fearful reaction and, invariably, to overmark it with their own spray.

The indoor perpetrator may be trying to increase confidence by surrounding him or herself with his or her own familiar smell. Hence cats which spray indoors may be trying to repair scent or security 'holes' in their own protective surroundings caused by change, the arrival of new objects or the addition of strange smells on new objects, cats or people, or the loss of a contributor to the communal smell that helps identify every member of the den. Spraying after cat flap installation probably occurs because the outdoors is then perceived as continuous with the indoors and the cat feels that the home therefore needs to be anointed to identify his occupancy and ensure that he encounters his own smell frequently. The cause should be identified if possible and the cat's exposure to any physical challenges controlled. Cat flaps should be boarded up to define den security, sprayed or middened areas cleaned as outlined in the 'House soiling' section and baited with dry food. The cat should never be punished either at the time or, worse, after the event, as this increases indoor insecurity and the need to mark. However, confining the cat to one room when unsupervised can create a new safe 'core' which needs no further identification by spraying. The best 'cores' are usually warm, draught-proof beds close to radiators or safe heat sources where a cat would often naturally choose to relax and sleep. In severe cases, the core can even initially take the form of such a bed in an indoor pen, with the cat only released to the secure surrounding room under supervision. Once the cat relaxes in the room and ceases to spray, his or her access through the house can be expanded gradually, one cleaned, baited room at a time, under the supervision of the owner at each stage. During treatment and perhaps for ever more, all local rival cats should be chased out of the garden to avoid upsetting or challenging the patient.

Protest sprayers should be ignored and the whole relationship between owner and cat restructured so the cat receives contact, food, affection, etc. only at the owner's initiation, as in the treatment of overdemanding or dominant dogs. This type of spraying may worsen initially before responding positively to treatment. Drug support is very much case-dependent but a tapered prescription by the veterinary surgeon of oral progestins or sedatives for three to four weeks may assist treatment.

House soiling

Inappropriate urination and defecation as acts of normal toileting, or as a result of nervousness, should first be distinguished from deliberate acts of marking by spraying or associative marking by urination and middening. Most cats instinctively tend to use loose substrate such as cat litter as their latrine when first venturing from the maternal nest, and learn by experimentation and observation of their mother that litter is a surface on and in which to excrete. Prior to this they are unable to excrete without physical stimulation from the mother. Initially this is carried out in the nest and the action enables the mother to clean all waste and prevent the kittens from soiling the nest. Later on in their development the mother may carry the kitten out of the nest and lick them to stimulate excretion, with the result that the majority of cats are taught early never to soil their own bed. In adulthood the house comes to be seen as an extension of the bed and a feeding lair. Excretion therefore normally takes place away from it, or is directed into a litter tray.

Poor maternal care can disrupt this learning process and occasionally kittens are weaned without becoming house-trained, especially some longhair strains. For others, medical or emotional trauma, especially during the cat's adolescence, decreases the security of home and an initial breakdown in hygiene may then continue long after the source of the problem has disappeared or been treated. Cats that are generally nervous or incompetent may excrete repeatedly indoors rather than venture outside, and the siting and nature of the litter tray, and type of litter offered can all affect toileting behaviour. Being offered food too close to the tray will deter many cats from using it, and if the tray is in a site that is too busy, open or otherwise vulnerable it may also cause cats to seek safer places. Some cats that are normally fastidious in their personal hygiene are reluctant to use soiled or damp trays, or to share with other cats. Trays may need to be cleaned more frequently or more trays may need to be provided. Certain compressed wood pellet litters appear to be less comfortable for cats to stand on, especially for cats who live permanently indoors and have more sensitive feet pads than cats whose pads are toughened up by an outdoor lifestyle. Litters which release deodorizing scents when damp have also been known to deter some cats from urinating in the tray, possibly because of the irritation they cause to the pads of the feet when damp. Inflammation and cornification of the pads should be looked for in such cases. Litters containing chlorophyll are also reported as being unattractive to some cats and so should be changed when problems arise.

Cats may also associate pain and discomfort with their tray if suffering from cystitis, feline urological syndrome or constipation. They may then seek alternative surfaces and continue to find carpets or beds more attractive as latrines. Other cats sometimes simply forget where the tray is, get 'caught short', or become arthritic or perhaps a little lazy in old age and need more trays or easier access to them.

Treatment

It is often worth trying a finer grain commercial litter or fine sterile sand, which cats, perhaps because of their semi-desert ancestry, seem to find more attractive than woodchip pellets or coarse-grain litters. The position of the tray should be checked, especially relative to the position of food bowls, and for security. Placing the tray in a corner or offering a covered tray, or, if the cat is unwilling to be enclosed, at least a tray with sides but no roof, may help to improve the security of the latrine. It should not be sited in active areas of the house or those disrupted by children or the family dog. The tray should be cleaned a little less frequently than previously to allow the smell of the cat's urine to accumulate, as this improves its identification and association as a latrine. As with dogs, the smell of urine may stimulate the cat to urinate, but the whole litter surface should not be allowed to get too dirty or too damp, as this may also deter the cat. Once a day cleansing per cat is usually adequate. For outdoor cats, up to 50 per cent soil from the garden can be added to the litter or sand. Transfer of the use of the litter tray completely outdoors over a period of two to three weeks can be achieved by moving the tray progressively nearer the door and then out on to the step and finally into the garden.

For serious cases, confinement in a small room for a few days may help to reduce the opportunity for mistakes. Confinement in a pen where a simple choice between tray and bed can be provided may ensure that any early learning is reinforced. The cat can steadily be allowed more freedom indoors, one room at a time, when able to target excretion into the tray. Previously soiled areas in the house must be thoroughly cleaned, but never with an agent that contains ammonia as this is a constituent of urine and may endorse the idea of the cleaned area being a latrine. Many proprietary agents/cleaners may only mask the smell to the human nose and not be effective for the cat. Instead, a warm solution of a biological detergent may be followed by a wipe or scrub down with surgical spirit or other alcohol. Certain dyes in fabrics may be affected and so should be checked first for colour fastness under this cleaning system. Cleaned areas should be thoroughly dry before the cat is allowed supervised access. Bowls of dry cat food (with the food glued to the bottom if necessary to prevent consumption) may act as a deterrent to toileting at cleaned sites for a few days and, longer term, the cat can also be fed its main meal at previously regularly soiled areas, provided, of course, that the family dog can be prevented from consuming it!

The cat should never be punished, even if 'caught in the act'. This makes cats more nervous and more likely to excrete in the house and even in the presence of the owner. Instead the cat should be calmly placed on its tray or outside the house and accompanied for reassurance. Timing of feeding can help to make faecal passage time more predictable in kittens and young cats and enable the cat to be put in the right place at the right time. Drug support is only usually helpful in cases of inappropriate toileting caused by nervousness. Prescription by the veterinary surgeon of oral sedatives or progestins for one to two weeks, in addition to management changes, may be beneficial in such cases.

Attachment/bonding difficulties

The critical time for socializing kittens to humans, other cats, dogs and a normal household environment, is between two and seven weeks of age. Most problems of nervousness and incompetence in adult cats would never have arisen had they been handled intensively during this period and exposed to a wide range of stimuli and experiences. Between four and twelve weeks (prior to completion of vaccination courses) they should be subjected to as complex and active a home environment as possible. Imprinting on humans in the few hours after birth probably also occurs through smell, and handling then may help produce a friendlier, tractable pet at weaning. Recent Anglo-Swiss research suggests that there may be two distinct character types in cats, one with a high requirement for social contact and one for which such contact may be tolerated but not seen as an essential feature of the quality of life. The latter group seem to have a higher requirement for social play and predatory activity than affectionate interactions. A cat may therefore need to live with other cats and be less competent socially on its own, or need to lead a solitary life and be less able to be sociable with other cats. It is suggested that, as a result, in its relations with human owners, a cat will either have a high requirement for physical contact and petting from the owner, or will never appreciate it, even if the owner is very insistent at trying to provide it. Cats more typical of this second category may well prove less rewarding as pets, especially to those owners seeking a very affectionate relationship based on physical contact with the cat. However, these categorizations take little account of temperament changes of cats in adulthood, or resulting from personality differences between one owner and the next. Nor does it consider the fact that most cats become far more affectionate towards their owners after, for example, intensive nursing following trauma or during illness. Furthermore, improvements may also result from trying to treat individual aggressive or nervous conditions. However, if a particular cat is rather shy and unresponsive to attempts to improve its self-confidence, it is usually comforting for the owner to know that the origins of the problem may be largely genetic and beyond their influence. Perhaps their cat is closer in character to his uncompromised and reactive ancestors whose speed or reaction to challenge defined their chances of survival.

Under-attachment

Cats perceived by their owners to be 'underattached' are often intolerant of the owner's proximity or approach, especially of handling, and fail to relax when held. Causes may include a lack of early socialization, over-enthusiasm on the part of the owner, trauma or necessary invasive handling during illness.

Treatment

Treatment involves increasing the bond with and dependence on the owners and a major feature of it is the feeding of frequent, small, attractive meals preceded by much vocal communication and encouragement for the cat to follow the owner for its food. Feeding at table level while attempting gentle handling along the cat's back only is the next step. Actions such as steadily increasing the frequency and intensity of handling, offering treats at other times and occupying favoured resting positions on the floor by the fire/radiator so that the cat comes to sit on the owner to gain access may all improve the cat's perception of the owner as rewarding. Owners should discontinue all efforts to chase the cat with a view to handling, especially if the cat seems to be more of the predatory and less affectionate character. The more owners try to initiate contact with this type of cat, the less time will actually be spent in contact with them. If, however, owners make themselves more attractive to their under-attached cats by offering food, titbits and toys or by lying passively in front of favourite resting places, such as the fire, and allowing the cat to initiate the interaction, then the total time spent with the cat will usually increase. In severe cases the cat can be penned for a short time to accustom it to close human presence, and though it may seem a little bizarre, owners should try to approach the cat head first to stimulate the greeting behaviour observed between friendly cats and introduce hands (perhaps otherwise viewed as threatening weapons) slowly afterwards.

It is essential that owners always respond positively with affectionate touch and a calm, gentle voice to any initiating gesture the cat may make in approaching them in the home, especially for cats which are allowed outdoors. Drug support is usually not necessary except with severely traumatized cats or those unhandled before about eight weeks, but tapered prescription of progestins may help.

Over-attachment

Typically such cats are of pedigree strains, kept on long after weaning in the breeder's home when they should have been developing new social attachments with other cats and people and developing their behavioural repertoire in response to new challenges in new environments. They are also more likely to be cats of any breed or crossbred which are kept permanently indoors and so are more dependent on their owners for stimulation and social contact. Over-attached cats may follow their owner constantly, perhaps crying regularly in an effort to engage them in physical contact. This is often the case in elderly cats if they are left alone and feel insecure during the night. Once the owners have responded to the cat's distress calls by getting up, perhaps because they suspect some medical problem in their pet, the cat, reassured, often simply settles back down to sleep. The over-attached cat may be agitated or nervous when isolated.

Often these cats demonstrate prolonged infantile behaviour when with their owners, such as sucking their clothes or skin. Owners may then feel guilty about rejecting the cat's affection or fear lack of contact if they do not respond. Over-attachment is often the fault of encouragement by the owner for close association where the cat fails to lose sucking and other nursing responses after weaning. It may also occur after intensive nursing during illness or during old age with its associated increased dependence on the owners.

Treatment

Treatment involves detachment by non-punishing rejection of the cat's advances, together with periodic physical separation and replacement by alternative forms of affectionate contact for short periods of time, initiated by the owner. Provision of novel objects helps the cat learn to explore. Aversion therapy using loud startling noises, or a sudden jet of water can be used in severe cases. Old cats can be offered a secure, warm bed in the owner's bedroom and will usually then remain reassured and quiet through the night without needing to cry out to gain immediate physical attention.

Nervousness, phobias, separation anxiety

Nervousness, phobias and separation anxieties are presented as a range of problems that vary from the cat failing to adapt to 'normal' household events such as noise and visitors to lack of confidence in individual family members to failure to cope when away from the owner and agoraphobia. Cats may be shy and fearful if not exposed to a range of experiences and handling between the ages of two and seven weeks. This is particularly likely if they are of the social play/ predatory type of character and they are pursued too frequently or handled too roughly at any stage. Behaviour includes becoming withdrawn and secretive, moving with a low crouching gait, reluctance to enter open space or go outdoors, loss of appetite or psychogenic vomiting in very severe cases. Low threshold flight reactions and defensive (fear) aggression may occur if the cat is unable to avoid the challenge. Cats with such fears may have suffered a lack of early experience or a trauma. Agoraphobia, for instance, may be caused by fear of attacks by cats outdoors. Indeed, agoraphobia is the only recognized genuine phobia encountered in cats. Old age and its associated loss of competence may also be a factor in the development of nervous conditions.

Treatment

The basic treatment consists of systematic desensitization involving controlled exposure to known problem stimuli in low but increasing doses while denying the opportunity to escape, so providing the possibility for habituation to the problem. With general nervousness/incompetence this is often best achieved by penning the cat indoors (or outdoors for agoraphobic cats if coupled with chasing other cats away) and forcing it to experience 'normal' household events such as the proximity of visitors, the family and other pets while protected. The cat may thus come to learn that their presence is not threatening and often does so very quickly. Frequent short meals should be offered by an increasing number of people, including visitors. Detachment from any one over-favoured member will encourage the cat to spread its loyalties to more people. Drug support is often helpful. Progestins can be administered, and oral sedatives during desensitization may help. Alternative medicines, such as certain homoeopathic treatments and Bach Flower Remedies, may also be helpful for longer term support of cats which are generally hyper-reactive to common stimuli and nervous.

Over-grooming and self-mutilation

Most cats groom their flanks or back when confused, immediately after some mild upset or when unable to avoid general threatening stimuli. The behaviour seems to have little effect on layering or quality of the coat but has a stress-relieving function. This may be mediated, as in social monkeys, by the release of opiates from grooming and repetitive self-interested behaviour patterns. The behaviour is usually harmless, but occasionally a cat will over-groom in response to continued 'stress', such as the presence of too many cats in the house, acquisition of a dog, isolation from its owner, physical punishment or harassment by the owner for other behaviours such as house soiling or in response to an emotional disturbance between family members. Grooming may progress to the point of breaking hairshafts and producing a balding appearance to the flanks, the base of the tail, on the abdominal area or on the legs. In severe cases of unresolved stress, or in particularly sensitive or incompetent individuals, the cat may actually pluck out large quantities of fur causing large bald patches. Often this is a secretive behaviour in its initial stages as the cat may feel more comfortable in the owner's presence and so refrain from the behaviour. In the later stages the cat may mutilate itself in their presence as well. This is an area where behaviourists and dermatologists are now conferring as it is thought that these reactions may also be triggered as a result of flea allergy and sensitivity to diet, and occasionally from allergy to household dust, but go far past the normal groom or scratch behaviour because of some underlying 'stress'.

I encounter actual self-mutilation extremely rarely but it is described as a common clinical condition in Canada by APBC member Professor Donal McKeown of Ontario Veterinary School. Such severe self-inflicted damage is usually directed at itchy infected plucked areas or, less explicably but typically, at the tail or mouth. In these cases the behaviour is usually manic and occurs in frequent or occasional episodes which may be self-reinforcing because of the euphoria engendered by the release of opiates, and the preferential status of the cat compared with relaxation or facing up to an unresolvable challenge. Many cases, such as sporadic clawing at the tongue presented approximately every six months by one Burmese cat in the UK, had no obvious clinical cause. Other obsessive compulsive disorders in cats have been described such as air-licking, prolonged staring, air batting, jaw snapping, pacing, head shaking, freezing, paw shaking and aggressive attacks at the tail or feet accompanied sometimes by vocalization.

Treatment

Any dermatosis or other possible medical cause such as flea sensitivity, atopy or dietary allergy, should initially be investigated and treated accordingly by the veterinary surgeon. Only when such potential causes have been ruled out, or established as contributory causes, can behaviour therapy be considered. Building the competence of a cat to cope alone by restructuring relations with the owner is required if separation anxiety is suspected. Stimulation with novel objects and situations, and controlled change of husbandry patterns can also be offered. Self-mutilation can sometimes be resolved in single cats by the acquisition of another cat. The use of an Elizabethan collar for a short time may also help healing and perhaps break any learned behaviour patterns. The cat may also be distracted with sudden movement, loud noises or jets of water during severe episodes of mutilation. Generally increased levels of contact initiated by the owner may also help to define relations better and to offer more security in the home without the cat becoming overdependent on the owner's presence.

The sedative diazepam may be given as immediate treatment to control severe episodes of self-mutilation and a lower dose during lifestyle modification lasting several weeks. Anti-convulsants, anti-depressant and anti-anxiety drugs have been employed with some success in cats and dogs in Canada, and morphine antagonists such as naloxone may inhibit the behaviour, perhaps by enabling the animal to feel the pain of its self-mutilation. The effectiveness of any drugs at treating such problems is believed to be influenced by the length of time the behaviour has been expressed and the presence and ability of the owner to control conflicts and stresses in the cat's lifestyle and home environment.

Aggression

Towards other cats

Aggression towards other cats may vary from occasional or frequent hissing or scuffling between two individuals in multi-cat households to serious physical attacks on all cats on sight, indoors or out. Despotic aggression, victimization and, most commonly, persistant intolerance of new feline arrivals to the household are all quite common and have given rise to the widely held belief that most cats are highly individualistic, territorial and intolerant of other cats. In fact the cat is one of the most socially 'elastic' mammals on earth and can adapt its social tolerance according to circumstances and the individual nature of other cats it encounters. Hence vast numbers of cats can be found living peaceably together in 'feral' colonies in a wide range of environments around the world, including some where food and shelter are in short supply and might understandably be thought to lead to increased social competition and levels of aggression. For individual pet cats which don't get on, aggression by one cat towards another may include violent physical attack, low threshold arousal in response to the sight of movement of other cats or a total lack of initial investigatory or greeting behaviour. The cat may also be generally hyperactive and territorial. Nape biting and mounting particularly of younger and unfamiliar or passive cats may also be observed. Aggression rarely seems to be a defensive reaction, but occasionally attack becomes a learned policy to avoid investigation by other cats.

Depending on its early experiences, a cat either may have an emotional need to share a home base with other cats or be more solitary. In the latter case, the cat may be able to tolerate other house cats, but never form close social ties based on mutual grooming and resource sharing. Causes may include individual dislike or intolerance of one or more individual cats or lack of social learning or contact with other cats when young. There may be marked territorial defence reactions with failure to recognize and respond to other cats' friendly or neutral reactions, which may compound the success of early assertive or rough play with siblings. Territorial defence reactions, mutual intolerance of un-neutered male cats, defence of kittening areas by fertile and oestrous queens or kitten defence by mothers are normal and expected forms of aggression and not regarded as treatable. Finally, medical conditions such as hyperthyroidism, brain lesions and diet sensitivity can also cause aggression problems, but diagnosis and treatment clearly lies in the hands of the veterinary surgeon.

Treatment

The approach has to be varied to suit the nature of a particular cat. For example, controlled frequent exposure to new arrivals by housing the original cat and the new arrival alternately in an individual pen can allow protected introductions. Distraction techniques can be employed, such as bringing cats together when feeding, modification of owner relations (especially with more 'rank conscious' oriental breeds) and instilling a hierarchy favouring the top cat in all greeting and play. In severe cases, a highly aroused protagonist can sometimes be deterred from attack by a spray of water from a plant mister and thereafter become more tolerant of the presence of another cat in the home, but such tactics are usually labour-intensive and may require repeated carefully managed introduction sessions to achieve lasting results. In severe cases, rehoming may be the only safe and kind option.

Drug support includes tapered doses of progestins, anti-androgenic injectables which may calm the aggressor (even if neutered), and progestins or sedatives. Some alternative treatments such as Bach Flower Remedies may help a traumatized victim relax more during controlled introductions.

Towards visitors and owners

Cats may attack people, grabbing them with their claws and biting, though this is rarely accompanied by vocalization. The behaviour is often sudden and unpredictable and may be triggered by sudden movement, such as passing feet, or occasionally by certain high-pitched sounds. Defensive aggression to prevent handling is often caused by a lack of either early socialization or gentle human contact. Predatory chasing of feet and other moving body targets, territorial defence, especially in narrow or confined areas (only seen so far in oriental breeds), hyper-excitement during play, dominant aggression towards people when vulnerable (e.g. lying or sitting down), occasionally food guarding and kitten defence against owners by nursing mothers have all been recorded. However the most difficult problem is that of redirected aggression by very territorial cats agitated by catching sight of rivals through a window.

Owners who approach to pacify their cat may inadvertently

stimulate an attack by their movement. 'Petting and biting syndrome' also occurs in many cats, but this is usually tolerated or avoided by the owner. Initially the cat accepts affection but it may then suddenly lash out, grab and bite the owner and then leap away to effect escape. The threshold of reaction is usually high and injury slight.

Treatment

The members of the family most at risk, such as those with jerky or unpredictable movement patterns, children and elderly relatives, for example, must be taken into account when devising treatment. Controlled exposure to normal family movements and activities can help to habituate the cat to them. However, stimulation in the form of another carefully introduced cat (preferably a kitten, which may be less threatening than a more socially, territorially or sexually competitive adult), together with the opportunity to go outdoors (free ranging where possible, or perhaps on a harness and lead in urban areas), frequent presentation of novel objects and concentrated play sessions (predatory chase, capture) of half an hour per day are all therapeutic. A moving target in the form of a ball or string will help an excited or frustrated cat release aggression more safely and facilitate owner intervention afterwards with less risk.

Diet sensitivity and its effects on feline behaviour are little understood but may be empirically investigated by withdrawing all canned food for two weeks and replacing it with fresh chicken/fish plus subsequent vitamin/mineral supplements. Modern complete dry diets can also be tried and the cat maintained on them in the long term if there is an improvement in its behaviour. This approach has been particularly valuable for those problem behaviours that have occurred after a change from a standard canned diet to an apparently better quality canned or 'cooked in foil' prepared diet.

Access to catnip toys should be denied during treatment to preclude concomitant excitability in sensitive cats. Drug support using progestins for three to six weeks may help and some 'alternative' medicinal approaches have produced excellent responses in easily agitated oriental breeds in particular. Burmese cats seem to respond especially well to certain Bach Flower Remedies.

Pica

This is the depraved ingestion of non-nutritional items. It is usually unexplained, although the eating of some house plants may be due to the desire to obtain roughage or minerals and vitamins. Occasional cases are reported of cats eating rubber and electric cables but the main problem concerns the ingestion of wool and other fabric. Wool eating was first documented in the 1950s and was thought to be limited to Siamese strains but a recent survey of 152 fabric eating cats by Dr John Bradshaw of Southampton University and myself shows the behaviour to be more widespread. Responses to the survey showed that fabric eating was presented by Siamese (55 per cent of responders) and Burmese (28 per cent) cats, occasionally by other oriental strains and, more rarely, by crossbred cats (11 per cent). Males are as likely as females to present the problem and the majority of both sexes in the survey were neutered. The typical age of onset for fabric eating is two to eight months. Most cats (93 per cent in the survey) start by consuming wool, perhaps attracted by the smell of lanolin, but later transfer to other fabrics. Sixty-four per cent also ate cotton and 54 per cent consumed synthetic fabric.

While some fabric eaters chew or eat material on a regular basis, others only do so sporadically. Many consume large amounts of material, such as woollen jumpers, cotton towels, underwear, furniture covers, etc., without apparent harm, although surgery is required in a few cases to clear gastric obstructions and impaction of material. Some have caused such damage to owner's property that they have been euthanased, but most owners of fabric eating cats seem remarkably tolerant of their cat's behaviour.

The exact cause of the behaviour is unknown but a variety of factors have been suggested, including genetic. It is suggested that wool eating is a largely inheritable trait that is rarely expressed and is caused by a physiologically based hyperactivity of the autonomic nervous system. Such neuronal disturbances could affect the control of the digestive tract and thereby produce unusual food cravings and inappropriate appetite stimulation, although the exact mechanism is unknown.

Another suggestion is that the desire to suck and knead wool and then other fabric is a continuing redirected form of suckling behaviour resulting from the failure of the cat to mature fully. Some cats grow out of the behaviour at maturity but others will eat all unattended fabric despite good nutrition and husbandry. The behaviour is usually more prevalent in cats housed permanently indoors. Fabric eating is sometimes secretive, but it is usually blatant and unaffected by punishment. While most cats will consume fabric at any time, some will take a woollen item to the food bowl and eat this alternately with their usual diet and only at mealtimes.

Some fabric eating cats have caused hundreds of pounds' worth of damage to designer jumpers, carpets and tweed-covered furniture! Fabric eating behaviour may sometimes be triggered by some form of stress, perhaps in the form of medical treatment or the introduction of another cat to the household. A significant number of cats in the survey first exhibited fabric eating within one month of acquisition. Insufficient handling of kittens before adoption or separation from the mother at too early an age may also lead to stress and trigger the behaviour. Only 15 per cent of cats in the survey were acquired before eight weeks of age, which is earlier than is often recommended for maximizing the prospect of as full an emotional development as possible, particularly for the later maturing pedigree oriental strains. Over half of the pedigree cats in the survey were acquired by their owners at or beyond the minimum age of twelve weeks recommended by the UK Governing Council of the Cat Fancy, suggesting that the age at which a kitten is taken from its mother may not be the only influence on the subsequent development of fabric eating.

Continuing infantile traits such as over-dependence on the physical presence of the owner can lead to separation anxiety when the owner departs and cause the cat to start to eat fabric. The behaviour may then be triggered during subsequent stressful experiences as a learned pattern, even in response to previously tolerated influences. The best hope for treatment of fabric eating probably rests with cases where the relationship between owner and cat can be modified so that the cat is made less dependent on its owners for emotional security and the need to eat fabric as a form of anxiety relieving displacement behaviour can be reduced.

Fabric eating also seems to form part of a prey catching/ingestion sequence otherwise usually unexpressed in the day-to-day repertoires of the pet cat fed prepared and often very easily digested food. Indeed, forty per cent of the cats reported in the survey had little or no access to the outdoors and hence restricted or no opportunity to develop exploratory and hunting behaviour, including ingestion of small prey.

Treatment

This currently involves a combination approach of social restructuring with the owner (see 'Over-attachment', page 139), increasing the level of stimulation for the cat through play, increased activity at home, opportunity to investigate novel stimuli and, where possible, the opportunity for indoor cats to lead an outdoor life. This can mean allowing free access or housing in a secure pen or accustoming the cat to being walked on a lead and harness.

Increasing the fibre content of the diet by offering a dry diet and/or gristly meat attached to large bones to increase food (prey) handling and ingestion time has also brought improvements in many and even a few total cures. Others have improved by being offered increased fibre in the form of bran or chopped undyed wool or tissue paper blended in with their usual wet canned diet. Such tactics presumably help because the higher fibre intake keeps the cat's stomach active and reduces any appetite-related motivations to fabric eating. Remote ambushes using aromatic taste deterrents such as eucalyptus oil or menthol applied to woollen clothes may be employed, though traditional deterrents using pepper or chilli powder seem only to broaden the cat's normal taste preferences! Remote aversion tactics using touch-sensitive cap exploders under clothes or under electric cables deliberately made available can deter some cats (this treatment should only be offered under the careful guidance of the veterinary surgeon or referred behaviourist and never used with nervous cats or those suffering from certain medical problems, such as most cardiac conditions) and some cats may be safely channelled into chewing only certain acceptable items provided at meal and resting times. Owners of such cats have found then that to preserve other clothes and household items they need to keep the cat supplied with a cheap supply of its favourite fabric. No drugs are as yet recognized as being helpful with treatment.

Suggested further reading
....

Association of Pet Behaviour Counsellors, Annual Report 1990, 257 Royal College Street, London, NW1.

M.W. Fox, *Understanding Your Cat*, Coward, McCann and Geoghegan Inc., New York, 1974.

B. L. and L. A. Hart *Canine and Feline Behavioural Therapy*, Lea & Febiger, Philadelphia, 1985.

U. A. Luescher, D. B. McKeown and J. Halip, 'Stereotypic or obsessive-compulsive disorders in dogs and cats,' *Advances in Companion Animal Behaviour, Veterinary Clinics in North America*, vol. 21, no.2, 1991.

P. F. Neville, *Claws . . . and purrs: Understanding the Two Sides of Your Cat*, Sidgwick & Jackson, Pan MacMillan, London, 1992.

——*Do Cats Need Shrinks?* Sidgwick & Jackson, Pan MacMillan, London, 1990.

——*Treatment of behaviour problems in cats*, Practice, vol. 13, no. 2, 1991, pp. 43–50.

P. F. Neville and J. W. S. Bradshaw, 'Unusual appetites,' *Bulletin of the Feline Advisory Bureau*, vol. 28, no. 1.

D. C. Turner, 'The ethology of the human–cat relationship,' *Schweiz Arch Tierheilk 133*, 1991.

D. C. Turner and P. Bateson *The Domestic Cat: The Biology of Its Behaviour*, Cambridge University Press, 1988.

D. C. Turner and M. K. Stammbach-Geering, *Owner Assessment and the Ethology of Human–Cat Relationships*, Veterinary Association Publications, London, 1990.

12 Pet Loss – Guilt, Grief and Coping
Anne McBride

Compared to only a few decades ago, many more of us are likely to be sharing our homes, daily lives and even our bed with a pet. The change in the way we regard animals and their role in our emotional life is clearly illustrated in the names we choose to call them. In the past, animals were given 'animal' names, like 'Nipper' and 'Tiger'. More recently, pets are given human names – amongst the most popular being 'Ben, 'Susie', 'Judy' and 'Sam'.

One consequence of the closer ties we have with our pets is the more profound reaction we have to their death. In general, the lifespan of our pets is short and most of us will outlive our companion. While the loss of a close human friend or family member is considered to be a major stress event and worthy of a display of grief, it is only recently that there has been wide acceptance of the idea that the loss of a companion animal can be just as stressful to the bereaved owner.

It is perhaps ironic that many of us, while fully accepting all the positive aspects of pet ownership, still find it difficult to acknowledge the naturalness of the grief process caused by the loss of a particular animal. The major exception is when the bereaved is a child, in which case we expect, and accept, the tears and the need to say goodbye with some gesture such as burying the pet in the garden. But any similar expression of grief from an adult at the loss of 'the dog/cat/budgie/rabbit/horse' tends to make us feel very awkward and uncomfortable, not knowing what to do or say in such circumstances. Likewise, the bereaved person himself is often surprised and acutely embarrassed by the intensity of his reaction to the loss of the pet. He may even feel guilty about his seemingly 'childish' reaction.

One reason for these contrasting feelings is the dichotomy of our relationship to animals, which is highlighted by the death of a pet. On the one hand is the thought that it is 'only an animal' that has died and, after all, we do not mourn for the cow as we tuck into our Sunday roast. On the other hand, this wasn't just any animal, this was 'Ben'. The conflict between thinking of 'Ben' as companion and 'Ben' as just an animal results in a problem for many of us when faced with a grieving owner. We do not know how to react to what we may consider is not 'real' grief. In truth, this attitude as to what constitutes 'real' grief is still shared by some doctors and professional bereavement counsellors, who may reject a plea for help from a distressed owner.

Until recently society has restricted its definition of legitimate or 'real' grief to the loss of a relative or close friend. Communities have developed a number of specific rituals for these circumstances. Though details vary between social groups, such rituals serve a common, dual function. First, they provide the bereaved person, for example a widow, with a strategy for coping with her loss. In this way, she can work through her grief and adjust her perception of the world and her role within it to her new situation. Second, formal procedures allow others to express their sympathy, if not their empathy, and give support to the bereaved. The rituals allow the sympathizer to offer condolences while remaining at an emotional distance consistent with their own ability to cope with death and with the widow's distress. The fact that funereal rites are practised all over the world among all peoples indicates their importance to the human psyche, both of the bereaved and of other members of the social group.

However, a period of mourning following the loss of a pet is not a new phenomenon. We know from tomb pictures that the Ancient Egyptians had dogs called such names as 'Akena' and 'Xabesu' which were obviously much-loved family pets. When a dog died it was embalmed, ritually mourned and buried in the town's special dog cemetery. More recently, since the beginning of the profession some two hundred years ago, vets have had to deal with the saddened pet owner. There is plenty of evidence of lamenting owners scattered throughout history as in Lord Byron's poem written in 1808 on the death of his Newfoundland dog, Boatswain. The poem, inscribed on the dog's gravestone, ends with the lines,

> To mark a friend's remains these stones arise;
> I never knew but one, – and here he lies.

What is new is the growing public acceptance that such grief is genuine and natural. However, as yet, bereaved pet owners do not receive the same level of support as those grieving for a fellow human being, and often are left to face their grief alone, and even in the face of ridicule from others. Understanding the grief process and being aware of the rituals available to help ease the pain are essential in helping both owner and concerned outsider to deal with the loss of a pet in a complete and satisfactory way.

Just as there is variety in the make-up of people so there is diversity in the reaction of different people to the death of a pet, and of the same person to the death of differing pets. We may simply have a feeling of sadness for a little while or we may find the death more difficult to cope with. This is particularly true when the animal has been special to us in some way. Perhaps it has been part of our life for many years – in the case of children often for the whole of their life. A pet may represent the ending of an era, a link with the past. I can remember, when in my twenties, being upset at the death of a very elderly rabbit because she was the last of my 'childhood pets'. For some the pet may have had significance for the owner as a link with a person who has died. For example, the death of the husband's dog can be very traumatic for a widow and lead her not only to grieve for the dog but also renew mourning for her husband.

Special relationships also develop with animals to whom we have given extra care. Animals we have raised as orphans, have rescued or have required some special attention and care during their life tend to mean 'more to us'. Likewise, animals which have been an integral part of important phases in our life can be much missed, as was the case with one owner of my acquaintance whose dog shared his bachelorhood and was present throughout his first marriage and the traumas of its break-up. As he put it, his dog represented 'a reliable source of love in an unreliable world'. In one way or another, we have made a substantial emotional investment in these 'special pets' and the price we pay for this is the deep pain we feel when they die.

The way we react to the death of a pet will also be influenced by our personality and our circumstances at the time. Those of us who are isolated from our fellows for some reason and rely on the pet as a source of love and companionship will naturally feel quite disconsolate at its death. It could be that the death occurs just when everything else seems to be going wrong and we are just not able to cope with this final blow. In all of these cases, the bereaved owner may react as if the loss was of a human friend.

Over the last few years much has been written about grief and the stages through which the mourner passes in his attempt to come to terms with his loss. While it is right to talk about 'stages' in mourning, it should be made clear that not everyone will follow the same pattern. The stages discussed below may overlap, melding together and replacing each other. Stages can reoccur, often when least expected. The terms 'stages' and 'process' of mourning imply a definite beginning and end, but in reality working through grief is much more complex and individualistic.

The first stage is characterized by shock, numbness and denial. This usually immediately follows the death of the pet but it can occur as an anticipatory reaction. For instance, when the veterinary surgeon tells the owner that the animal has an incurable illness or, simply, that the effects of old age mean that there is no longer the quality of life we would wish for our friend. On hearing such news the owner may suddenly feel quite cold or feel as if someone has just hit them in the stomach. As C. S. Lewis put it, 'No one ever told me that grief felt so like fear. I am not afraid, but the sensation is like being afraid.'

At the time of first hearing of the gravity of our pet's condition, we may just not be able to cope. Many vets will recognize the situation where they 'break the bad news' only to have the owner react as if nothing has been said, perhaps asking the vet to carry out some irrelevant task such as cleaning the dog's ears. This reaction is perfectly normal and can be considered adaptive. The numbing of our consciousness to the total awfulness of the news is a defensive reaction. It enables us to deal with the gravity of the situation in manageable chunks. In some cases, it may take repeated consultations with the vet before the reality of the fading animal's condition is accepted.

It could be that the vet is only warning us that our pet is reaching its natural end or we may be being asked to consider taking the matter into our own hands. Obviously, this is not the time to make a decision about euthanasia, and the sensitive vet will make clear that he understands our sense of disbelief and our feeling of impending loss. A rushed decision can lead to deep feelings of guilt and regret for the owner, and possibly resentment towards the veterinary surgeon.

If the animal is not suffering, then the wise vet will ask the owners to take a day or two to come to terms with the prospect of euthanasia and to say goodbye in whatever manner seems appropriate, be it a last walk or the preparation of some special meal. The Pyrenean Mountain Dog that accompanied me through the first fourteen years of my life had, for medical reasons, only been allowed to drink boiled water. During his last twenty-four hours he had been unable to stand unaided and was to be euthanased at home the next morning. The family's goodbye gesture consisted of a large bowl of cool fresh water and, in my case, sitting up with him all night. Just to reiterate the point about aspects of the grief process reoccurring, I am not totally surprised that I feel tearful when writing about this episode . . . several decades after the event.

Sometimes, as when a pet has been involved in a road traffic accident, there is no time to overcome the initial shock, no time to say goodbye. The animal may have been killed outright or been so badly injured that immediate euthanasia was the kindest procedure. In this case, the owner may be very shocked and numbed.

Whatever the severity of the shock reaction it is often a great comfort to know that there are people around who care and understand. I have heard owners talk warmly of the veterinary nurse or receptionist who took a few moments to talk to the stunned owner before they left the surgery, some even providing a cup of tea. Others speak of the letter of sympathy they received from the veterinary surgeon following the death of their pet. Such gestures are important as many owners are unable to express their feelings easily for fear of receiving an unsympathetic response. Even within a family, everyone may be so busy trying to cheer each other up that no one has the chance to show how they really feel. Talking over the situation and providing an opportunity for everyone to be upset, without shame, is far more constructive.

It's all right to cry.

Only when the initial shock is over should any decision be made regarding the details of the practicalities surrounding the death of the pet. Such issues may include when and where euthanasia is to occur, at home or in the surgery, and who is to be present. Some members of the family may wish to be with the pet during its final moments, others may feel that they would rather have a few moments alone with the body immediately afterwards. It is important that everyone has a chance to express their wishes and as far as possible to have them fulfilled. Each of us needs to have the opportunity to say goodbye.

A vet told me recently of a dog that had been brought in after a road traffic accident. The dog was badly injured and could not be saved, indeed its injuries were rather gruesome. The owners requested that the vet dispose of the body but asked if their children could come and see their dog for the last time. Many of us may well have been appalled by the thought of letting the children see the broken remains of their friend and persuaded the parents that this would not be a good idea. This vet, however, made an arrangement for the family to come over later in the day. She then spent time roughly sewing up the dog, shampooing it to remove the blood, drying it with a hair dryer and arranging the body so that it looked peaceful. This vet certainly had the welfare of both her animal clients and their owners at heart.

In this particular case the vet was asked to dispose of the dog's body. This is one of several options available to the owner and careful thought should be given as to which is most suitable. Some pets are buried in the garden, but this is often not an option for city dwellers. A friend in the country may oblige or you may wish to bury your pet in a pet cemetery. An alternative to burial is cremation, after which the ashes are returned to you in a small casket. This could be buried or the contents scattered in a special place in the garden or on a walk. Some prefer to keep their pet's ashes indoors, and I know of at least one couple who take the cremated remains of their canine friend on their caravan holidays. The Victorians often had their favourite dog preserved and mounted in a glass case. More recently Roy Rogers had his famous partner, the palomino horse Trigger, treated similarly. While some of us may feel this is a little too much for our taste, that doesn't mean it isn't right for others.

Many owners find comfort in marking the final resting place of their pet with a stone, plaque or plant. Even if there is no grave, as when the vet is responsible for the disposal of the body, it is perfectly reasonable to mark the loss of the pet by burying some memento such as the pet's collar, planting a shrub or erecting some sort of memorial. Children and adults may feel the need to do something active to help them express their emotions and work through their grief. This may be the drawing of a picture, writing a poem or an obituary to be printed in an obituary column in one of the dog or other pet magazines. At present, pet obituary columns are few and far between, but I predict that they will become more common.

I have already talked about the shock and denial which occurs after the death of a pet. Denial can continue for some time and catch us quite unawares. We may wake up on a sunny morning thinking what a lovely day to take 'Ben' for a walk, only to remember that he was euthanased a few days earlier. Or we may find ourselves getting out of our chair to open the tin of cat food, just as we have done every day for the last seventeen years, only to recall with a jolt of pain that now it is not needed. Even finding oneself walking into the pet shop or to the pet food aisle of the supermarket is not so unusual. These behaviours are not silly, they are perfectly natural and again resemble the activities of someone who has lost a close human partner. There are numerous reports of people automatically setting a place at the table for the dead person, or finding themselves waiting by the phone for a call which can never come.

Another stage in the grief process is that of guilt. We tend to relive the events surrounding the death, blaming ourselves for actions we did or did not take, chastising ourselves with thoughts of, 'If only . . .' 'If only I'd taken him to the vet earlier,' 'If only I had had him on a lead.' This attempt to turn back the clock and somehow change the course of events is a natural reaction and does fade with time and with acceptance of the fact that our pet is no longer with us.

The guilt felt by the pet owner is quite understandable when we realize that, for all the time the pet was with us, we were responsible for its happiness and physical well-being. Our responsibilities also include the way in which our pet ends its life. Even though we consult with our vet and feel in our hearts that euthanasing our old or ill pet is the right thing to do, we may still feel that we are 'killers' and have betrayed the trust placed in us over the years. Making that final decision between life and death is a formidable task and it is not surprising that we continue to feel bad long after the event. However, if we did not make the decision rashly and knew the animal was suffering, then we can derive solace from the fact that our pet died a peaceful, painless death; the kind of death we would wish for anyone.

Sometimes we have not been in a position to make such a decision and a perfectly healthy life suddenly has ended. An unexpected death tends to produce even more self-accusation and suffering in the guilt-ridden owner. The pet may have met with a fatal accident which the owner feels he could have avoided: the horse that broke out of the paddock, the dog that rushed out of the house, the rabbit or guinea pig left out in its pen at night to the delight of the local fox. The unexpected death of a healthy pet can also result from the demands of the law. Dogs involved in worrying sheep, attacking people or other dogs can all pay the penalty with their life, much to the heartache of their owners who may have been relatively responsible. The intensity of guilt and self-reproach in the owners of such pets is understandable and to be expected. There is often little comfort available, other than the passage of time and being able to learn from the mistake and prevent it being repeated in the future.

In addition to blaming ourselves for the death of the pet, we often direct blame on to others. Understandably, we are unlikely to think favourably of the person who tells us, 'Stop being silly, after all it was only a cat.' Likewise, we may have good reason to feel angry with someone if they were partly, or wholly responsible for the pet's demise through some wilful or thoughtless action or even neglect. A recent study of nearly 1,000 bereaved owners (Lee and Lee, 1992) showed that 80 per cent of owners whose animals had died in a road traffic accident were angry with the 'cause of the accident', presumably the driver of the vehicle.

Our anger can also be directed at innocent people. We may be short-tempered and irritable with friends and family and even other pet owners – comparing the virtues of our dead pet with the supposed vices of their living one. The vet is another person who often comes in for a dose of vitriolic feeling from the bereaved owner, especially if the pet died of an acute illness. We may berate him for not doing enough, not being sufficiently skilful to have saved our pet, even though, deep down, we know this not to be the case. Even God gets allotted a share of the blame and anger. It is important that those on the receiving end realize that this is all part of the grief process, another stage to be passed through, maybe more than once.

Bereaved people often suffer spells of depression, they may feel tired yet are unable to sleep properly. Conversely, they may be very restless, not being able to settle, feeling something is missing (which it is) and unconsciously searching for it. Occasionally, this unconscious yearning to find the lost pet surfaces, as when we dream about our companion. We may even feel their presence with us around the house or on a favourite walk. Sometimes we may even think we see them. Such illusions are quite common and perfectly normal. It could be that it is not the individual pet for which we are searching, rather it is the 'animalness' which is now missing from our lives. I say this not to belittle the importance of that particular pet, rather to put the process into context. Namely, this yearning tends to cease when another animal is brought into the household. Likewise, it can be less intense for owners who have more than one pet.

These initial potent stages of grief last a few days, weeks or months. They are succeeded by a period of remembering, the beginning of recovery. The owner will often want to talk about his deceased pet, share stories about the good, as well as the more recent bad events. As listeners, we may get impatient, especially if the same anecdote is repeated several times. Again, it is worth remembering that this is all part of the process and that just by listening we can be helping the owner heal themselves.

Bereaved owners are often told to 'get another one' and so they should, when, and only when, they are ready. For some, being without an animal is in itself traumatic and they may get a new companion immediately. These people, and I am one, do not feel they are replacing an individual, rather the newcomer is replacing the missing 'animalness'. Indeed, it is very important that the new pet, which may be of a different breed or even species, is not brought home as a substitute. It will have its own character and will form its own special bond with us, if we allow it to do so. If we treat it as a substitute, it will never live up to our memory of its predecessor and we will never be able to have a satisfying relationship with it.

Other people need more time before they are ready to make the emotional investment in a new companion animal. There is no rush and we should not feel pressured, nor pressure anyone into 'getting a new one'. Sometimes people are too frightened of the possibility of being hurt again and cannot face having another animal. If they can be helped to come to terms with their loss properly, perhaps with the assistance of a pet bereavement counsellor, then they may well be able to rediscover the pleasures of pet ownership. Elderly people, too, often feel unable to replace their pet. Perhaps they think they cannot cope with another dog or a young animal, either because they feel physically incapable or because they are worried what would happen if they died before the pet. It could be that a different species which is less demanding, such as a budgie, or an older, well-adjusted animal needing a home is the answer.

Even from this short description of the grief process, we can see that it is not straightforward. Different emotions come into play: shock, denial, guilt, depression and anger. They mix inside us like some cocktail, and we have little control over either the ingredients or proportions. Even when we think we are completely over the loss of our friend, the smallest memory can bring it all flooding back, a fact I can testify to when writing this chapter.

Some losses are easier to cope with, some more difficult. We may think that those who are likely to suffer more are people who we consider to be more lonely in general, but we shouldn't be complacent. Even within a happy family situation some members may have built a deeper bond with the dead animal than others and be vulnerable to suffering intense grief. And grief over the loss of a pet can be very intense. Lee and Lee (1992) report eight cases in a two-year period of suicide committed by owners following the death of the pet. A staggering number, and probably not the actual total of pet-related suicides. Even more unexpected was the fact that seven of those suicides were men. This alone underlines the need for us all to be more sympathetic, understanding and tolerant of the needs of the bereaved pet owner.

We tend to think of bereavement as only relating to death, but really it is to do with loss. For the pet owner, the grief process can be set in train if a pet goes missing. If not found, there is no opportunity to say goodbye, no ending to the mourning. It is a difficult situation and the owner who is never reunited with their pet may always suffer sadness and, perhaps, guilt. Another form of loss is when owner and animal are parted for some reason. This may be because age and frailty require the owner to move into sheltered accommodation where pets are not allowed. A change of job, or losing one's home may necessitate giving up one's pet to another's care. Assistance animals, such as guide dogs, often have to be rehomed when they are retired because the owner's circumstances do not allow them to be kept purely as a pet (Nicholson and Kemp-Wheeler, 1992). Even knowing the animal is being well looked after rarely ameliorates the feelings of loss, and the sense of guilt at betraying our animal friend.

There are two other groups whose reactions to pet death need to be considered. First is our veterinary surgeon, who, if you remember, we may well have yelled at in our anguish. Vets are in the business because they have a respect for animals. At graduation they took a vow to do the best they can for the animals in their care, yet most of those working with companion animals will euthanase several animals a week, every week of their working life. Some of these animals may be beyond help, others may be perfectly healthy and being destroyed at the behest of the owner or the law. Little research has been done on the effects of this aspect of their work. That which has, such as the study by Fogle and Abrahamson (1990) shows that vets suffer short- and long-term emotional effects. Comments by vets in this study showed that they were touched by the death of an animal: 'Every animal I destroy leaves a little bit of me dying,' 'Knowing it was important to someone as a friend, companion, or worker makes it worse. All the fun and enjoyment it brought to someone is gone and that makes me feel bad inside.' We hope that our vet understood our anger, and he probably did, but there is no harm in our appreciating that euthanasing animals is no joyride for him or her either.

The final area of pet bereavement I would like to touch on is the reaction of other pets in the household. The species we tend to have as companion animals, for instance cats, dogs, birds and horses, are social animals. It is this aspect of their behaviour that enables us to form an understanding, or bond, with them. But we are not the only creatures they form bonds with, they do so with other animals in the household. There are numerous anecdotes, and I can relate a few myself, of cats calling for their lost companion or dogs moping around the house and going off their food for a few days. It doesn't even have to be a companion of the same species. Dogs can 'mourn' cats and vice versa, and there are many stories of pets pining for their dead owners. Konrad Lorenz (1963), one of the founding fathers of the study of animal behaviour, wrote of the separation of a greylag goose from its mate that, 'All the objective observable characteristics of the goose's behaviour on losing its mate are roughly identical with human grief.' Likewise Dr John Bowlby (1961), the renowned investigator of child psychology, on reviewing the literature of species as diverse as the jackdaw, dog and chimpanzee, concluded that, 'Members of lower species protest at the loss of a loved object and do all in their power to seek and recover it; hostility, externally directed, is frequent; withdrawal, rejection of a potential new object, apathy and restlessness are the rule.' Sounds familiar, doesn't it? As with the grieving owner, patience and understanding can help other pets come to terms with the loss of their companion.

We have our pets to care for, to love and to enjoy the many pleasures they give us in return. The death of a pet means that particular relationship can only continue in our hearts and it is natural that we should react to that loss. But, once we have mourned and found a permanent, secure place for our pet in our memories, we may feel that the most fitting tribute to our dead companion is to build a new relationship with another animal friend.

To Scott
(A Collie, for nine years our friend)

Old friend, your place is empty now. No more
Shall we obey your imperious deep-mouthed call
That begged the instant freedom of our hall.
We shall not trace your foot-fall on the floor
Nor hear your urgent paws upon the door.
The loud-thumped tail that welcomed one and all,
The volleyed bark that nightly would appal
Our tim'rous errand boys – these things are o'er.

But always yours shall be a household name,

And other dogs must list' your storied fame;
So gallant and so courteous, Scott, you were,
Mighty abroad, at home most debonair.
Now God who made you will not count it blame
That we commend your spirit to His care.

W. M. Letts, 1916

References

J. Bowlby, 'Process of mourning', *International Journal of Psychoanalysis*, vol. 44, 1961.

B. Fogle and D. Abrahamson, 'Pet loss: a survey of the attitudes and feelings of practising veterinarians', *Anthrozoos*, vol. 3, 1990.

M. and L. Lee, 'British attitudes towards pet loss: role of the veterinarian when a pet dies', Abstracts of Sixth International Conference on Human Animal Interactions, 1992.

C. S. Lewis, *A Grief Observed*, Faber, London, 1961.

K. Lorenz, *On Aggression*, Methuen, London, 1963.

C. Murray-Parkes, *Bereavement: Studies of Grief in Adult Life*, Pelican, London, 1981.

J. Nicholson and S. Kemp-Wheeler, 'The End of a partnership: guide dog owners' reactions to the loss of a working guide dog', Abstracts of Sixth International Conference on Human Animal Interactions, 1992.

Suggested further reading

M. and L. Lee, *Absent Friend*, Henson Ltd, London, 1992.

SCAS, *Death of an Animal Friend*, Straight Line Publishing, Glasgow, 1990.

Further Help

The following material by APBC members is available from the Honorary Secretary, APBC, 257 Royal College Street, London, NW1. Prices quoted include postage and packing.

For help with problems, write to the same address and enclose a stamped, addressed envelope.

Books

John Fisher, *Dogwise: The Natural Way to Train Your Dog*. Pbk £12.49.
——*Think Dog! An Owner's Guide to Canine Psychology*. Hbk £12.45.
——*Why Does My Dog . . . ?* Hbk £15.49.
Sarah Heath, *Why Does My Cat . . . ?* (1993 publication). Hbk £15.49 (est.).
Anne McBride, *Rabbits and Hares*. Hbk £6.95.
Peter Neville, *Claws and Purrs*. Pbk £11.49.
——*Do Cats Need Shrinks?* Pbk £8.50.
——*Do Dogs Need Shrinks?* Pbk £11.49.
with Russell Jones, cartoonist, *Memoirs of an Animal Shrink*. Pbk £5.50.
Valerie O'Farrell, *Problem Dog: Behaviour and Misbehaviour*. Hbk £11.99, pbk £7.49.
Katie Patmore, *So Your Children Want a Dog?* Pbk £7.49.
John Rogerson, *Be Your Dog's Best Friend* (cartoons by McLachlan). Pbk £4.99.
——*Training Your Dog*. Hbk £17.49.
——*Understanding Your Dog*. Hbk £14.49.
——*Your Dog, Its Development Behaviour and Training*. Hbk £15.49.

Booklets

David Appleby, 'How to Have a Happy Puppy'. £2.75.
——'The Good Behaviour Guide' (Dog Help). £2.75.

Erica Peachey, 'Running Puppy Classes: Points to Consider for New and Established Classes'. £4.50.

Videos (VHS only)

All prices include VAT.

John Fisher, *Training the Dog in the Human Pack*. 60 mins, £14.49.

Peter Neville, *Cool for Cats*. A programme to entertain cats living permanently indoors. £9.99.

——*Cool for Dogs*. Relaxation programme for dogs left alone. Helps treat and prevent separation anxiety. 60 mins, £9.99.

——and John Tandy, *Feline Behaviour Therapy*. 58 mins, £20.00.

and Bradley Viner, *A Dog's Life and How to Enjoy It*. 68 mins, £13.99.

——and Bradley Viner, *All You Need to Know About Cats*. 65 mins, £13.99.

John Rogerson, *An Introduction to Tracking*. 60 mins, £14.49.

——*The Dominant Dog*. 30 mins, £17.99.

Products

All prices include VAT.

Dog Training Discs by John Fisher. £7.50.

Walk-Easy Headcollar for dogs by Dog Help. State size: small, medium, large or extra large. £5.50.

Walkee Headcollar/Training and Exercise Lead by John Fisher. One adjustable size suits all dogs of all ages. £5.50.

INDEX

Abrantes, Dr Roger vi, 121, 126
Abyssinian cats: behaviour problems 131
acepromazine maleate (ACP) 85-6, 87
adopting dogs *see* rehomed dogs
aggressive behaviour:
  in cats:
    towards other cats 131, *132*, 143-4
    towards humans *132*, 138, 144-5
  in dogs *19*, *23*, 41, 43-5
    causes of 47-8
    and diet 46
    and dominant behaviour *see* dominance
    drugs and 46, 84, 87
    fear and 1-2, 30, 42, 45, 50, *see also* fear
    fighting and 115
    over food 48, 49
    genetic predisposition to 14
    and greeting rituals *22, 23*
    punishment and 47-8, 50-1, 117-18
    with siblings 14, 15, *15*, 17, 19
    see also phobias; rages, psychopathic
agoraphobia (cats) *132*, 140-1
alopecia (cats) 73
amitriptyline 85
anal glands, infected (dogs) 118
anorexia (dogs): drugs for 84-5
anthropomorphism 110
anxiety *see* fear; separation anxiety;
    nervousness
Appleby, David vi, 6
Association of Pet Behaviour Counsellors
    (APBC) 11
Atenolol 87, 88

Bach Flower Remedies 141, 144, 145
Bailey, Gwen vi-vii
barking: preventing in puppies:
  when left 38
  at tradesmen 37
  at visitors 36
  see also separation anxiety
behavioural problems:
  and medical causes 4, 10-11, 118-19
  and lack of training 10
  vets and 4-5, 7, 9, 10, 11, 119
  see also drugs, use of *and specific problems*
benzodiazepines 85
beta-blockers 71, 87-8
biting, play (dogs) 37-8, 54, 56
  of siblings 15, *15*, 16, 17, 19
Blue Cross 93
Bowlby, Dr John 158
Boxers 64
Bradshaw, Dr John 146
breeders, dog: and socialization of puppies 6,
    30-1, 33, 34-6

burial of pets 153-4
Burmese cats: behavioural problems in 131,
    132, 142, 145, 146-8

cars: accustoming puppies to 39
  leaving dogs in 66
  nausea in 86
  territorial behaviour in 30
cat flaps: and spraying 134, 135
cats: breed differences 5-6
  and humans 130-1, 138-40
  and puppies 26, 37
catteries, cats in 73-4
chewing (dogs): of paws 72, 91
  when left 38, 64, 70, 75, 98
children: accustoming puppies to 34, 36, 39
  treating for dog phobia 81
choke chains (dogs) 52, 59, 117
collar, first (dogs) 59
countryside, accustoming puppy to 39

death of pets:
  and buying another 156
  depression after 155-7
  guilt and 154-5
  and mourning 149-52
  reaction of other pets to 66, 158
  vets and 152-3, 155, 157
  see also burial
defecation *see* house-training problems
destructiveness: and noise 77-9
  when left alone 38, 64, 70
dexamphetamine 85
diazepam 86-7, 142
diet and behaviour: of cats 145, 147-8
  of dogs 46
dominance (in dogs) 12, 20-2, *23*, 96, 97, 100,
    127-8
  techniques to reduce 45, 49, 94-5, 97, 122-3
'Down!' (command) 125-6
drugs, use of 11 (for dogs), 133
  (for cats). *See specific problem*
Dunbar, Dr Ian vii, xi, 7

epileptic seizures *see* seizures

fabric eating (by cats) 146-7
  treatment 147-8
fear: in cats *see* nervousness
  in dogs *23*, 43-4, 83-4
    and lack of socialization 24, 34
    of men 34-5, 48, 49-50
    in puppies 12-14, 63-4, 102-3
    when being socialized 40
    see also aggressive behaviour; phobias;
        submission
feeding: guarding of food and (of dogs) 48, 49

162